AF559891

INTRODUCTORY OPERATION RESEARCH

By

S.C. Sharma

DISCOVERY PUBLISHING HOUSE PVT. LTD.
NEW DELHI-110 002

First Published - 2006

Reprinted - 2018

ISBN: 978-81-8356-091-7

Published by:

DISCOVERY PUBLISHING HOUSE PVT. LTD.

4383/4B, Ansari Road, Darya Ganj

New Delhi-110 002 (India)

Phone: +91-11-23279245, 23253475, 43596065

E-mail: discoverypublishinghouse@gmail.com

orderdphbooks@gmail.com

web: www.discoverypublishinggroup.com

Printed at:

Infinity Imaging Systems

Delhi

Preface

This book on Operation Research has been specially written to meet the requirements of the M.Sc., M.Com. and M.B.A. students for all Indian Universities.

The subject matter has been discussed in such a simple way that the students will find no difficulty to understand it. The proof of various theorems and examples has been given with minute details. Each chapter of this book contains complete theory and fairly large number of solved examples, sufficient problems have also been selected from various universities examination papers.

In the preparation of this book large number of books and research papers have been consulted. So no authenticity is claimed.

The author express has gratitude to Mr. Wasan and staff of M/s Discovery Publishing House for their whole hearted co-operation in the publication of this book.

The author tried hard to be accurate and upto date in statement and realises the impressibility of completely avoiding error's. Therefore the author will greatly appropriate having his authentic called to any questionable statement.

Author

Contents

1

Introduction to Operation Research

OPERATION RESEARCH AND DECISION-MAKING

Decision is, in fact, a pivotal word in managing. It is not only the headache of management, rather all of us make decisions. We daily decide about minor 10 major issues. We choose to be engineers, doctors, etc., a vital decision which is going to affect us throughout our lives. We choose to purchase at a particular shop—a decision of relatively minor importance.

In fact, there is a scope of large scale improvement. The essential characteristics of ail decisions are as follows:

(i) Objectives,

(ii) Alternatives,

(iii) Influencing factors (constraints).

Once these characteristics are known, one can think of improving the characteristics so as to improve upon the decision itself.

Let us consider a situation in which a decision has been taken to see a particular movie and the problem is to decide She conveyance. Three alternatives are available: rickshaw, autorickshaw and a local bus.

In the first level of decision-making, autorickshaw is chosen as the mode of conveyance just by intuition. *i.e.*, it is decided at random. Evidently, it is a highly emotional and qualitative way of decision-making.

In the second level of decision-making, the three conveyances are compared and it is decided qualitatively that autorickshaw will be preferred since, though a little costlier, it is tune-saving and more comfortable.

In the third level of decision-making, the three alternatives are compared and it is suggested that autorickshaw will be chosen, as it will be taking only 1/3rd time than an ordinary rickshaw and shall be only 10% costlier while more comfortable. The local bus is rejected since it would not reach the theatre in time at all.

Though, outcome of all these decisions is the same, still we can judge the quality of each decision. We may brand the first decision as 'bad' since it is highly emotional, while we may call the second decision as 'good' since it is scientific though qualitative. The third decision is doubtlessly the best as it is scientific and quantitative.

It is this scientific quantification used in operation research , which helps management to make better decisions. Thus, in operation research , the essential features of decisions, namely, objectives, alternatives and influencing factors are expressed in terms of scientific quantifications or mathematical equations. This gives rise to certain mathematical relations, termed as a whole as mathematical model. Thus the essence of operation research is such mathematical models. For different situations different models are used and this process is continuing since World War II when the term operation research was coined. However, with the advance of science and technology, decision-making in business and industry has become highly complex and extremely difficult. The decision-maker is not only faced with a large number of interacting variables, which at times do not lend themselves to neat quantitative treatment but also finds them too numerous and dynamic. Above all he has to lake into consideration the actions of the competitors over which he has no control. This complexity of decision-making made the decision-makers look for various aids in decision-making. It is in these situations that operations research comes to our help. The managers today make full use of the operation research techniques in various functional areas. It has been realised beyond doubt that intuition alone has no place in decision-making since such a decision becomes highly questionable when it involves the choice among several alternatives. operation research provides the management much needed tools for improving the various decisions.

Operations research (OR) is a quantitative aid to decision-making offers the decision-maker a method of evaluating every possible alternative (act or course of action) by using various techniques to know the potential outcomes. This is not to say, however, that management decision-making is simply about the application of operations research techniques.

Generally, while solving a real-life problem, the decision-maker must examine it both from quantitative as well as qualitative perspective. Information about the problem from both these perspectives needs to be brought together and assessed in the context of the problem. Based on some mix of the two sources of information, a decision should be taken by the decision-maker. For example, consider a problem of an investor considering investments in three alternatives: Stock Market, Real Estate and Bank Deposit. To suggest an acceptable solution, we need to consider certain quantitative factors and information to be examined in the light of the problem. For instance, factors like financial ratios from the balance sheets of several companies whose stocks are under consideration; real estate companies' cash flows and rates of return for investment in property; and how much the investment will be worth in future when deposited at a bank at a given interest rate for a certain number of years, need to be examined. However, before reaching a conclusion, certain other qualitative factors, such as weather, state and central policies, new technology, the political situation, etc., need to be considered. These factors may be difficult to quantify.

Operations research (OR) must maintain stride with above said trends in the global age. Criticism has periodically been levelled at the failure of OR to adequately meet the needs of business and industry. Often the reasons for failure are behavioural in nature, and significant research on implementation process has been carried out. Focussing on implementation, however, presumes that the definition, analysis, modelling, and solution phases of an OR project have been adequately performed. One can easily hypothesize that among the reasons for implementation failure is the lack of creative problem solving abilities by the analyst. This book introduces the set of operations research techniques for decision-makers to help them make rational and effective decisions by having a basic knowledge of mathematics and statistics, as well as the use of computer software needed for computational purposes.

The evaluation of each alternative is extremely difficult or time consuming for two reasons: First, the amount and complexity of information that must be processed, second the number of alternative solutions could be so large that a decision-maker simply cannot evaluate all of them to select an appropriate one. For these reasons when there is a lack of qualitative factors, decision-makers increasingly turn to quantitative methods and use computers to arrive at the optimal solution to problems involving a large number of alternatives. The study of these methods and how decision-makers use them in the decision process is the essence of operations research approach.

HISTORY OF THE OPERATION RESEARCH

It is generally agreed that operations research came into existence as a discipline during World War-II when there was a critical need to manage scarce resources.

Pre-World War-II

No science has ever been born on a specific day. Operations research is no exception. Its roots are as old as science and society. Though, the roots of operation research extend to even early 1800s, it was in 1885 when Frederick: W. Taylor emphasised the application of scientific analysis to methods of production, that the real start took place. Taylor conducted experiments in connection with a simple shovel. His aim was to find that weight load of ore moved by shovel which would result in maximum of ore moved with minimum of fatigue. After many experiments with varying weights, he obtained the optimum weight load, which though much higher than that commonly used, provided maximum movement of ore during a day. For a "first-class man" the proper load turned out to be 20 pounds. Since the density of ores differs greatly, a shovel was designed for each ore so as to assume the proper weight when the shovel was correctly filled. Productivity rose substantially after this change.

Another man of early scientific management era was Henry L. Gantt. Most job-scheduling methods at that time were rather haphazard. A job, for instance, may be processed on a machine without trouble but then wait for days for acceptance by the next machine. Gantt mapped each job from machine to machine, minimizing every delay. Now, with the Gantt procedure it is possible to plan machine loadings months in advance and full quote delivery dates accurately.

A.K. Erlang in 1917, Danish mathematician published his work on the problem of congestion of telephone traffic. The difficulty was that during busy periods, telephone operators were unable to handle the calls the moment they were made resulting in delayed calls. A few years after its appearance, his work was accepted by the British Post Office as the basis for calculating circuit facilities. The formulae he developed on waiting time are of fundamental importance to the theory of telephone traffic.

F.W. Harris, published his work on the area of inventory control in 1915. During the 1930s, H.C. Levinson, an American astronomer, applied scientific analysis to problems of merchandising. His work included scientific study of customers' buying habits, response to advertising and relations of environment to the type of article sold.

This revolution, most of the industries were small scale, employing only a handful of men. The advent of machine tools—the replacement of man by machine as a source of power and improved means of transportation and communication resulted in fast flourishing industry. It became increasingly difficult for a single man to perform all the managerial function (of planning, sale, purchase, production, etc.). Consequently, a division of management function look place. Managers of production, marketing, finance, personnel, research and development etc. began to appear. With further industrial growth further subdivisions of management functions took place. The production department was sub-divided into sections like maintenance quality control procurement, production planning, etc.

This industrial development, brought with it, a new type of problems called executive type problems. These problems are a direct consequence of functional division of labour in an organization. In an organization, each functional unit (department or section) performs a part of the whole job and for its successful working, develops its own objectives. These objectives, though in the best interest of the individual department, may not be in the best interest of the organization as a whole. In fact, these objectives of individual departments may be inconsistent and clashing with each other. Consider, for example, the attitudes of he various departments of a business organization towards the inventory policy. The production department wants to have maximum production, associated with the lowest possible cost. This can be achieved by producing only one item continuously. Thus, it is interested in long, uninterrupted production runs, because such runs minimise set-up and clean-up costs. Thus, it prefers to have a large inventory in relatively few product lines.

The marketing department also wants a large but diverse inventory so that a customer may be provided immediate delivery over a wide variety of products. It would also like to have a flexible production policy so as to meet special demands at a short notice.

The finance department wants to minimize inventory so as to minimize the unproductive capital investments 'tied up' in it. It also believes that inventories should rise and fall with rise and fall in company's sales.

The personnel department wants to hire good labour and to retain it. This is possible only when goods are produced continuously for inventory during slack periods also. In other words, it is interested in maintaining a constant production level resulting in large inventory.

To set an inventory policy which serves the interest of the organization be a whole and not that of any individual department is an executive-type problem, which can be satisfactorily solved by the application of operation research techniques. The decision which is in the best interest of the organization as a whole is called *Optimal (optimum) decision* and the one in the best interest of an individual department is called sub-optimal decision.

World War-II

During World War II, England military management called on a team of scientists to study the strategic and tactical problems of air and land defence. This learn was under the direction of Professor P.M.S. Blackett of

Univerisity of Manchester and a former naval officer. "Blackett's circus", as the group was called, included three physiologists, two mathematical physicists, one astrophysicist, one army officer, one surveyor, one general physicist and two mathematicians. Many of the problems were of executive type. The objective was to find out the most effective allocation of limited military resources to the various military operations and to the activities within each operation. The application included the effective use of newly invented radar, allocation of British Air Force Planes to missions and the determination of best patterns for searching submarines. This group of scientists formed the first operation research team.

The encouraging results of these efforts led to the formation of more such teams in British armed services and the use of such scientific teams soon spread to Western Allies—the United States, Canada and France. Thus, though this science of operations research originated in England, the United States soon took the lead. In United States these operation research teams helped in developing strategies for mining operations, inventing new flight patterns and planning of sea mines.

World War-III

The success of military teams attracted the attention of industrial managers' who were seeking solutions to their problems. Industrial operations research in U.K. and U.S.A. developed along different lines. In U.K., the critical economic situation required drastic increase in production efficiency and creation of new markets. Nationalisation of a few key industries further increasing the potential field, for operation research. Consequently, operation research soon spread from military to government, industrial, social and economic planning.

In USA the situation was different. Impressed by its dramatic success in U.K., defence operations research in U.S.A. was increased. Most of the war experienced operation research workers remained in military services. Industrial executives did not call for much help because they were reluming to the peace-time situation and many of them believed that it was merely a new application of an old technique. Operations research has been known by a variety of names in that country such as operational analysis, operations evaluation, systems analysis, systems evaluation, systems research and management science.

The progress of industrial operations research in U.S.A. was due to advent of Second industrial Revolution which resulted in automation—the replacement of man by machine as a *source of control.* The new revolution began around 1940s when electronic computers became commercially available. These electronic brains possessed tremendous computational speed and information storage. But for these digital computers. Operations research with its complex computational problems could not have achieved its promising place in all kinds of operational environments.

In 1950, operation research was introduced as a subject for academic study in American Universities. Since then this subject has been gaining ever increasing importance for the students of Mathematics, Statistics, Commerce, Economics, Management and Engineering. To increase the impact of operations research, the Operations Research Society of America was formed in 1950. In 1953, the Institute of Management Sciences (IMS) was established. Other countries followed suit and in 1959 International Federation of operation research Societies was established and in many countries International Journals on operation research began to appear. Some of them (in English) are:

1. Operations Research,
2. Opsearch,
3. Operational Research Quarterly,
4. Management Science,
5. Transportation Science,
6. Mathematics of Operations Research,
7. International Journal of Game Theory, etc.

Today, the impact of operations research can be felt in many areas. This is shown by the ever increasing number of educational institutions offering this subject at degree level. The fast increasing number of management consulting firms speaks of the popularity of the subject. Of late, operation research activities have spread to diverse fields such as hospitals, libraries, city planning, transportation systems, crime investigation, etc. Some of the Indian organisations using operation research techniques Indian Airlines, Railways, Defence Organizations, Fertilizer Corporation of India, Delhi Cloth Mills, Tata Iron & Steel Co., etc.

However, a particular model and technique of operation research can be traced back to much earlier times. The term 'operations research' was coined as a result of research on military operations during this war. Since the war involved strategic and tactical problems which were greatly complicated, to expect adequate solutions from individuals or specialists in a single discipline was unrealistic. Therefore, groups of individuals who collectively were considered specialists in mathematics, economics, statistics and probability theory, engineering, behavioural, and physical science were formed as special units within the armed forces to deal with strategic and tactical problems of various military operations.

Such groups were first formed by the British Air Force and later, the American armed forces formed similar groups. One of the groups in Britain came to be known as Blackett's Circus. This group, under the leadership of Prof. P.M.S. Blackett was attached to the Radar Operational Research unit and was assigned the problem of analysing the coordination of radar equipment at gun sites. The efforts of such groups, especially in the area of radar detection are considered vital in Britain winning the air battle. Following the success of this group, such a mixed-team approach was also adopted in other allied nations.

After the war ended, scientists who had been active in the military operation research groups made efforts to apply the operations research approach to civilian problems, related to business, industry, research and development, etc. There are three important factors behind the rapid development in the use of operations research approach.

(i) Many operations researchers continued their research after war. Consequently, some important advancements were made in various operations research techniques. A key person in the post-war development of operation research was George B Dantzig. In 1947, he developed the concept of linear programming and its solution by a method known as simplex method. Besides linear programming, many other techniques of operation research, such as statistical quality control, dynamic programming, queuing theory and inventory theory were well developed before the end of the 1950.

(ii) Analytic power was made available by high-speed computers. The use of computers made it possible to apply many operation research techniques for practical decision analysis.

(iii) The economic and industrial boom after World War-II resulted in continuous mechanization, automation, decentralization of operations and division of management functions. This industrialization also resulted in complex managerial problems, and therefore application of operations research to managerial decision-making became popular.

During the 1950s, there was substantial progress in the application of operation research techniques for civilian activities along with a great interest in the professional development and education in operation research. Many colleges and universities introduced OR in their curricula. They were generally schools of engineering, public administration, business management, applied mathematics, economics, computer science, etc. Today, however, service organizations such as banks, hospitals, libraries, airlines, railways, etc., recognize the usefulness of operation research in improving efficiency. In

1948, an operation research club was formed in England which later changed its name to the operational Research Society of UK. Its journal, operation research Quarterly, first appeared in 1950. The operations Research Society of America (ORSA) was founded in 1952 and its journal. Operations Research was first published in 1953. In the same year, The Institute of Management Sciences (TIMS) was founded as an international society to identify, extend and unify scientific knowledge pertaining to management. Its journal. Management Science, first appeared in 1954.

In India, operations research came into existence in 1949 when an operation research unit was established at Regional Research Laboratory, Hyderabad for planning and organizing research. At the same time Prof. R.S. Verma also set up an operation research team at Defence Science Laboratory to solve problems of store, purchase and planning. In 1953, Prof. P.C. Mahalanobis established an operation research team in the Indian Statistical Institute, Calcutta to solve problems related to national planning and survey. The operation research Society of India (ORSI) was founded in 1957 and started publishing its journal Opsearch from 1964. In the same year, India along with Japan became a member of the International Federation of operational Research Societies (IFORS) with its headquarters in London. The other members of IFORS were U.K., U.S.A., France and West Germany.

A year later, project scheduling techniques: Program Evaluation and Review Technique (PERT) and Critical Path Method (CPM) were developed as efficient tools for scheduling and monitoring lengthy, complex and expensive projects of that time. By the 1960s operation research groups were formed in several organizations. Educational and professional development programmes were expanded at all levels and certain firms specializing in decision analysis were also formed. The American Institute for Decision Sciences came into existence in 1967. It was formed to promote, develop and apply quantitative approach to functional and behavioural problems of administration. It started publishing a journal, Decision Science, in 1970.

Because of OR'S multi-disciplinary character and application in varied fields, it has a bright future, provided people devoted to OR study can help meet the needs of society. Some of the problems in the area of hospital management, energy conservation, environmental pollution, etc., have been solved by operation research specialists and this is an indication that operation research can also contribute towards the improvement in the social life and areas of global need. However, in-order to make the future of operation research brighter, its specialists have to make good use of the opportunities available to them.

SOME DEFINITIONS OF THE OPERATION RESEARCH

Because of the wide scope of application of operations research, giving a precise definition is difficult. However, a few definitions of operation research are given below:

- Operation research is the application of the methods of science to complex problems in the direction and management of large systems of men, machines, materials and money in industry, business, government and defence. The distinctive approach is to develop a scientific model of the system incorporating measurements of factors such as chance and risk, with which to predict and compare the outcomes of alternative decisions, strategies or controls. The purpose is to help management in determining its policy and actions scientifically. —*Operational Research Society, U.K.*
- Operation research is concerned with scientifically deciding how to best design and operate man-machine systems usually requiring the allocation of scarce resources. *Operations Research Society, America*

Apart from being lengthy, the definition given by operational Research Society of U.K., has been criticized because it emphasizes complex problems and large systems, leaving the reader with the impression that it is a highly technical approach suitable only to large organizations. The definition of operation research Society of America contains an important reference to the allocation of scarce resources. The keywords used in the above definitions are scientific approach, scarce resources, system and model. The British definition contains no reference to optimization, while the American definition has no reference to the word, best. A few other definitions which are commonly used and are widely acceptable are as follows:

- Operation research is the systematic application of quantitative methods, techniques and tools to the analysis of problems involving the operation of systems. —*Daellenbach and George, 1978*
- Operation research is essentially a collection of mathematical techniques and tools which in conjunction with a systems approach, are applied to solve practical decision problems of an economic or engineering nature. —*Daellenbach and George, 1978*

These two definitions imply another view of operation research as being the collection of models and methods which have developed largely independent of one another.

- Operation research utilizes the planned approach (updated scientific method) and an interdisciplinary team in order to represent complex functional relationships as mathematical models for the purpose of providing a quantitative basis for decision-making and uncovering new problems for quantitative analysis.

—Thierauf and Klekamp, 1975

- This new decision-making field has been characterized by the use of scientific knowledge through interdisciplinary team effort for the purpose of determining the best utilization of limited resources.

—H A Taha, 1976

These two definitions refer to the interdisciplinary nature of OR. However, there is nothing which can stop one person from considering several aspects of the problem under consideration. One of the best definitions is as follows.

- Operation research, in the most general sense, can be characterized as the application of scientific methods, techniques and tools, to problems involving the operations of a system so as to provide those in control of the operations with optimum solutions to the problems.

—Churchman, Ackoffand Arnoff, 1957

- Operation research is the art of winning wars without actually fighting them. *—Auther dark*

This definition refers to the military origin of the subject where team of experts were not actually participating in military operations for winning the war but providing advisory and intellectual support for initiating strategic military actions.

- Operation research is the art of finding bad answers to problems to which otherwise worse answers are given. *—T.L. Saaty, 1958*

This definition refers operations research as a technique for selecting the best course of action out of several courses of action to reach a desirable solution of the problem.

A few other definitions of Operation research are as follows:

- Operation research has been described as a method, an approach, a set of techniques, a team activity, a combination of many disciplines, an extension of particular disciplines (mathematics, engineering, economics), a new discipline, a vocation, even a religion. It is perhaps some of all these things. *—S.L. Cook, 1977*
- Operation research may be described as a scientific approach to decision-making that involves the operations of organizational system.

—F.S. Hiller and G.J. Lieberman, 1980

- Operation research is a scientific method of providing executive departments with a quantitative basis for decisions under their control.

—P.M. Morse and G.E. Kimball

- Operations research is applied decision theory. It uses scientific, mathematical, or logical means to attempt to cope with the problems that confront the executive, when he tries to achieve a through-going rationality in dealing with his decision problems.
  —D.W. Miller and M.K. Stan
- Operations research is a scientific approach to problem-solving for executive management. *—H.M. Wagner*

As the discipline of operations research grew, numerous names were given to it, such as Operations Analysis, Systems Analysis, Decision Analysis, Management Science, Quantitative Analysis, Decision Science. However, each of these emphasizes the quantitative approach to the analysis and solution of management problems.

SCIENTIFIC METHOD IN OPERATION RESEARCH

The most important feature of operations research is the use of the scientific method and building of decision models. There are three phases of the scientific method on which operation research approach is based.

Research Phase

This phase is the largest and longest among other two phases. However, the remaining two are also equally important as they provide the basis for a scientific method. This phase utilizes:

(i) observations and data collection for a better understanding of the problem,

(ii) formulation of hypothesis and model,

(iii) observation and experimentation to test the hypothesis on the basis of additional data,

(iv) analysis of the available information and verification of the hypothesis using pre-established measures of desirability,

(v) prediction of various results from the hypothesis, and

(iv) generalization of the result and consideration of alternative methods.

Action Phase

This phase consists of making recommendations for implementing the decision by an individual who is in the position to implement results. This individual must be aware of the environment in which the problem occurred, objective, assumptions and omissions of the model of the problem.

Judgement Phase

This phase includes:

(i) identification of the real-life problem,

(ii) selection of an appropriate objective and the values of various variables related to this objective,

(iii) application of the appropriate scale of measurement, *i.e.,* deciding the measures of effectiveness (desirability), and

(iv) formulation of an appropriate model of the problem, abstracting the essential information, so that a solution to the decision-maker's goals can be obtained.

FEATURES OF OPERATIONS RESEARCH APPROACH

The features of operation research approach to any decision and control problem can be summarized as under.

Methodological Approach

Operations research is the application of scientific methods, techniques and tools to problems involving the operations of systems so as to provide those in control of operations with optimum solutions to the problems (Churchman *et al.*). The scientific method consists of observing and defining the problem; formulating and testing the hypothesis; and analysing the results of the test. The data so obtained is then used to decide whether the hypothesis should be accepted or not. If the hypothesis is accepted, the results should be implemented. Otherwise an alternative hypothesis has to be formulated.

Objectivistic Approach

An operations research approach seeks to obtain an optimal solution to the problem under analysis. For this, a measure of desirability (or effectiveness) is defined, based on the objective(s) of the organization. A measure of desirability so defined is then used to compare alternative courses of action with respect to their outcomes.

Interdisciplinary Approach

Interdisciplinary teamwork is essential because while attempting to solve a complex management problem, one person may not have the complete knowledge of all its aspects (such as economic, social, political, psychological, engineering, etc.). This means we should not expect a desirable solution to managerial problems. Therefore, a team of individuals specializing in mathematics, statistics, economics, engineering, computer science, psychology, etc., can be organized so that each aspect of the problem could be analysed by a particular specialist in that field in order to arrive at an appropriate and desirable solution of the problem. However, there are certain problem situations which may be analysed even by one individual.

Wholistic Approach

While arriving at a decision, an operations research team examines the relative importance of all conflicting and multiple objectives and the validity of claims of various departments of the organization from the perspective of the whole organization.

SCOPE OF OPERATIONS RESEARCH

In the field of industrial management, there is a chain of problems starting from the purchase of raw materials to the dispatch of finished goods. The management is interested in having an overall view of the method of optimizing profits. In order to take decision on scientific basis, operation research team will have to consider various alternative methods of producing the goods and the return in each case. Operation research study should also point out the possible changes in the overall structure like installation of a new machine, introduction of more automation, etc. Operation research has been successfully applied in industry in the fields of production, blending, product mix inventory control, demand forecast sale and purchase, transportation, repair and maintenance, scheduling and sequencing planning, scheduling and control of projects and scores of other associated areas.

Operation research has a wide scope for application in defence operations. In modern warfare the defence operations are carried out by a number of different agencies, namely airforce, army and navy The activities performed by each of them can be further divided into sub-activities *viz.* operations, intelligence, administration, training and the like. There is thus a need to coordinate the various activities involved in order to arrive at optimum strategy and to achieve consistent goals. Operations research, conducted by team of experts from all the associated fields, can be quite helpful to achieve the desired results.

In *both developing and developed economies*, operation research approach is equally applicable. In developing economies, there is a great scope of developing an operation research approach towards planning. The basic problem is to orient the planning so that there is maximum growth of per capita income in the shortest possible time, by taking into consideration the national goals and restrictions imposed by the country. The basic problem in roost of the countries in Asia and Africa is to remove poverty and hunger as quickly as possible. There is, therefore, a great scope for economists, statisticians, administrators, technicians, politicians and agriculture experts working together to solve this problem with an operation research approach.

Operation research approach needs to be equally developed in agriculture sector on national or international basis. With population explosion and consequent shortage of food, every country is facing the problem of optimum allocation of land to various crops in accordance with climatic conditions and available facilities. The problem of optimal distribution of water from the various water resources is faced by each developing country and a good amount of scientific work can be done in this direction.

Operation research approach is equally applicable to big and small organizations. For example, whenever a departmental store faces a problem like employing additional sales girls, purchasing an additional van, etc., tehniques of operation research can be applied to minimize cost and maximize benefit for each such decision.

Operation research methods can also be applied in big hospitals to reduce waiting time of out-door patients and to solve the administrative problems.

Monte Carlo Methods can be applied m the area of transport to regulate train arrivals and their running Limes. Queuing theory can be applied to minimize congestion and passengers' waiting time.

Operation research is directly applicable to business and society. For instance, it is increasingly being applied in L.I.C. offices to decides the premium rates of various policies. It has also been extensively used in petroleum, paper, chemical, metal processing, aircraft, rubber, transport and distribution, mining and textile industrials.

Thus, we find that Operation research has a diversified and wide scope in the social, economic and industrial problems of today.

GENERAL METHODS FOR SOLVING OPERATIONS RESEARCH MODELS

In general, the following three methods are used for solving operation research models. In all these methods, values of decision variables are obtained that optimize the given objective function (a measure of effectiveness).

Analytica! (or Deductive) Method

In this method, classical optimization techniques such as calculus, finite difference and graphs are used for solving an operation research model. In this case, we have a general solution specified by a symbol and we can obtain the optimal solution in a non-iterative manner. For example, in the inventory model, in order to calculate economic order quantity (optimal order size), the analytical method requires that the first derivative of the mathematical expression

$$TC = (D/Q)\ C_0 + (Q/2)\ C_h$$

where TC = total variable inventory cost

D = annual demand

C_0 = ordering cost per order

C_h = carrying cost per time period

Q = size of an order,

be taken and equated to zero as the first step towards identifying optimum value of,

$$EOQ(Q) = \sqrt{2DC_0IC_h}.$$

This is because of the concept of maxima and minima for optimality. Here, it may be noted that the validity of the conclusion depends only on the validity of the mathematical approach that is being used.

Monte Carlo Method

This method is based upon the idea of experimenting on a mathematical model by inserting into the model specific values of decision variables at different points of time and under different conditions and then observing their effect on the criterion chosen for variables. In this method, random samples of specified random variables are drawn to know what is happening to the system for a selected period of time under different conditions. The random samples form a probability distribution, that represents the real-life system and from this probability distribution, the value of the desired random variable can be estimated.

Numerical (or Iterative) Method

When analytical methods fail to obtain the solution of a particular problem due to its complexity in terms of constraints or number of variables, a numerical (or iterative) method is used to get the solution. In this method, instead of solving the problem directly, a general algorithm is applied to obtain a specific numerical solution.

The numerical method starts with a solution obtained by trial and error, and a set of rules for improving it towards optimality. The solution so obtained is then replaced by the improved solution and the process of getting an improved solution is repeated until such improvement is not possible or the cost of further calculation cannot be justified.

MODELS AND MODELLING IN OPERATION RESEARCH

Both simple and complex systems can easily be studied by concentrating on some portion or key features instead of concentrating on every detail of it. This approximation or abstraction, maintaining only the essential elements of the system, which may be constructed in various forms by establishing relationships among specified variables and parameters of the system, is called a model. In general, models attempt to describe the essence of a situation or activity by abstracting from reality so the decision-maker can study the relationship among relevant variables in isolation. Models do not attempt to duplicate reality in all aspects, but for models that do reveal nothing. Hence, models do not, and cannot, represent every aspect of reality because of the innumerable and changing characteristics of the real life problems to be represented. Instead, they are limited approximation of reality. For example, to study the flow of material through a factory, a scaled diagram on paper showing the factory floor, position of equipment, tools, and workers can be constructed. It would not be necessary to give such details as the colour of machines, the heights of the workers, or the temperature of the building. In other words, for a model to be effective, it must be representative of those aspects of reality that are being investigated and have a major impact on the decision situation.

It appears that a model is constructed to analyse and understand the given system for the purpose of improving its performance. The reliability of the solution obtained from a model depends on the validity of the model in representing the system under study. A model, however, allows the opportunity to examine the behavioural changes of a system without disturbing the on-going operations.

The key to model building lies in abstracting only the relevant variables that affect the criteria of the measures of performance of the given system and expressing the relationship in a suitable form. However, a model should be as simple as possible so as to give the desired result. But oversimplifying the problem can lead to a poor-decision. Model enrichment is accomplished through the process of changing constants into variables, adding variables, relaxing linear and other assumptions, and including randomness.

Classification Based on Function or Purpose

Models based on the purpose of their utility include the following types.

Descriptive Models

Descriptive models simply describe some aspects of a situation, based on observation, survey, questionnaire results or other available data of a situation and do not predict or recommend anything.

Examples of such models are: Organization chart, plant layout diagram, block diagram representing an algorithm or method for solving a problem, etc.

Predictive Models

These models indicate 'if this occurs, then that will follow'. They relate dependent and independent variables and permit trying out, 'what if questions. In other words, these models are used to predict the outcomes due to a given set of alternatives for the problem. These models do not have an objective function as a part of the model to evaluate decision alternatives.

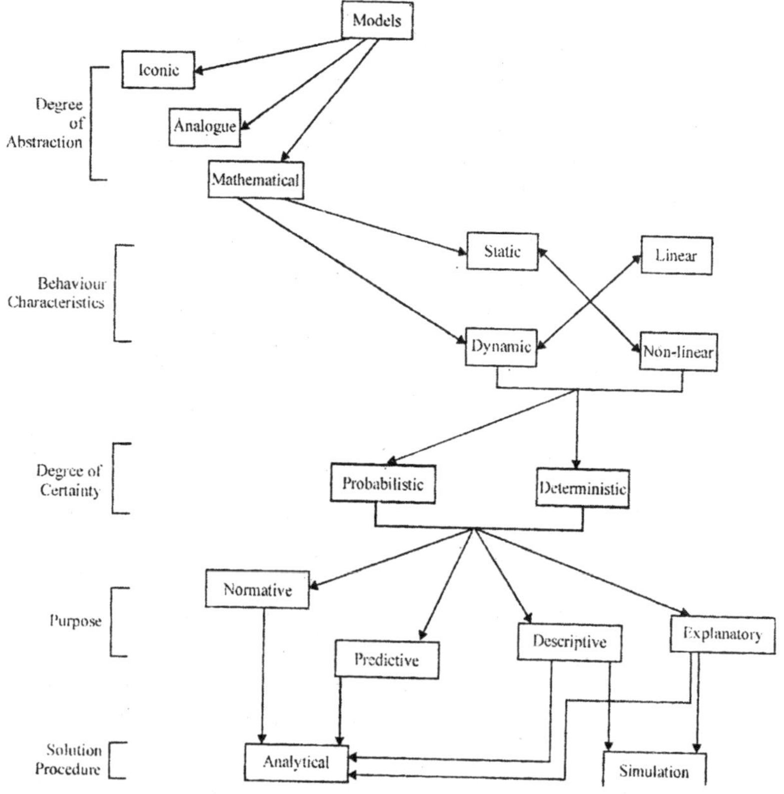

Fig. 1.1 : Classification of Models.

For example, S = a + bA + cI is a model that describes how the sale (S) of a product changes with a change in advertising expenditure (A) and disposable personal income (I). Here, a, b and c are parameters whose values must be estimated. Thus, having estimated the values of a, b and c, the value of advertising expenditure (A) can be adjusted for a given value of I to study impact of advertising on sales. In these models, however, one does not attempt to choose the best decision alternative, but can only have an idea about each alternative available to him.

Normative (or Optimization) Models

These models provide the 'best' or 'optimal' solution to problems subject to certain limitations on the use of resources. These models provide recommended courses of action. For example, in mathematical programming, models are formulated for optimizing the given objective function, subject to restrictions on resources in the context of the problem under consideration and non-negativity of variables. These models are also called prescriptive models because they prescribe what the decision-maker ought to do.

Classification Based on Structure

Physical Models

These models provide a physical appearance of the real object under study either reduced in size or scaled up. Physical models are useful only in design problems because they are easy to observe, build and describe. For example, in the aircraft industry, scale models of the proposed new aircraft are built and tested in wind tunnels to record the stresses experienced by the air frame. Since these models cannot be manipulated and are not very useful for prediction, problems such as portfolio selection, media selection, production scheduling, etc., cannot be analysed with a physical model. Physical models are classified into the following two categories.

(i) *Iconic Models:* Iconic models retain some of the physical properties and characteristics of the system they represent. An iconic model is either in an idealized form or is a scaled version of the system. In other words, such models represent the system as it is by scaling it up or down (*i.e.,* enlarging or reducing the size).

Examples of iconic models are, blueprints of a home, maps, globes, photographs, drawings, air planes, trains, etc.

Iconic models are simple to conceive, specific and concrete. An iconic model is used to describe the characteristics of the system rather than being explanatory. This means that such models are used to represent a static event and characteristics which are not used in determining or predicting effects due to certain changes in the actual system. For example, colour of an atom does not play any vital role in the scientific study of its structure. Similarly, type of engine in a car has no role to play in the study of parking problem.

(ii) *Analogue Models:* These models represent a system by a set of properties different from those of the original system and does not resemble it physically. After the problem is solved, the solution is reinterpreted in terms of the original system.

For example, the organizational chart represents the state of formal relationships existing between members of the organization. Maps in different colours may represent water, desert and other geographical features. Graphs of time series, stock-market changes, frequency curves, etc., may be used to represent quantitative relationships between any two properties and predict how a change in one property effects the other. These models are less specific and concrete but are easier to manipulate and are more general than iconic models.

Symbolic Models

These models use symbols (letters, numbers) and functions to represent variables and their relationships to describe the properties of the system. These models are also used to represent relationships which can be represented in a physical form. Symbolic models can be classified into two categories.

(i) *Verbal Models:* These models describe a situation in written or spoken language. Written sentences, books, etc., are examples of a verbal model.

(ii) *Mathematical Models:* These models involve the use of mathematical symbols, letters, numbers and mathematical operators (+, –, ÷, ×) to represent relationships among various variables of the system to describe its properties or behaviour. The solution to such models is then obtained by applying suitable mathematical technique.

The relationship among velocity, distance and acceleration is an example of a mathematical model. In accounting, the cost-volume-profit model is also an example of a mathematical model. Symbolic models are precise and abstract and can be analysed and manipulated by using laws of mathematics. The models are more explanatory rather than descriptive.

Classification Based on Time Reference

Static Models

Static models represent a system at some specified time and do not account for changes over time. For example, an inventory model can be developed and solved to determine an economic order quantity for the next period assuming that the demand in planning period would remain the same as that for today.

Dynamic Models

In a dynamic model, time is considered as one of the variables and allows the impact of changes due to change in time. Thus, a sequence of interrelated decisions over a period of time are made to select the optimal course of action to optimize the given objective. Dynamic programming is an example of a dynamic model.

Classification Based on Method of Solution or Quantification

Simulation Models

These models also have a mathematical structure but are not solved by applying mathematical techniques to get a solution. Instead, a simulation model is essentially a computer-assisted experimentation on a mathematical structure of a real-life problem in order to describe and evaluate its behaviour under certain assumptions over a period of time.

Simulation models are more flexible than mathematical ones and therefore, can be used to represent a complex system which otherwise cannot be represented mathematically. These models do not provide general solution like those of mathematical models.

Heuristic Models

These models employ some sets of rules which, though perhaps not optimal, do facilitate solutions of problems when applied in a consistent manner.

Analytical Models

These models have a specific mathematical structure and thus can be solved by known analytical or mathematical techniques. Any optimization model (which requires maximization or minimization of an objective function) is an analytical model.

Classification Based on Degree of Certainty

Deterministic Models

If all the parameters, constants and functional relationships are assumed to be known with certainty when the decision is made, the model is said to be deterministic. Thus, in such a case, the outcome associated with a particular course of action is known. That is, for a specific set of input values, there is a uniquely determined output which represents the solution of the model under conditions of certainty. The results of the models assume single value. Linear programming models are examples of deterministic models.

Probabilistic (Stochastic) Models

Models in which at least one parameter or decision variable is a random variable are called probabilistic (or stochastic) models. Since at least one decision variable is random, therefore, an independent variable which is the function of dependent variable(s) will also be random. This means consequences or payoff due to certain changes in the independent variable cannot be predicted with certainty. However, it is possible to predict a pattern of values of both the variables by their probability distribution.

Insurance against risk of fire, accidents, sickness, etc., are examples where the pattern of events is studied in the form of a probability distribution.

OPERATIONS RESEARCH IN INDIA

An operation research unit under Professor P.C. Mahalonobis was established in 1953 in the Indian Statistical Institute, Calcutta to apply OR methods in national planning and survey. Operations Research Society of India was formed in 1957 and its first conference was held in Delhi in 1959. It was felt that there existed the need of producing well-trained operations researchers who could tackle practical problems. It was also decided to bring out a journal on operations research, the first volume of which came out in 1963 with the name 'Opsearch'. Some other Indian journals promoting the cause of operations research are 'Industrial Engineering and Management', 'Materials Management Journal of India', 'Defence Science Journal', 'Journal of the Indian Society of Statistics and Operations Research' (ISSOR), 'Pure and Applied Mathematika Sciences' (PAMS), etc.

Professor Mahalonobis made the first important application of operation research in India in preparing the draft of the Second Five Year Plan. The Draft Plan frame is still the most scientifically formulated plan bearing programmes of massive economic development of India. It was estimated that India could become self-sufficient in food merely by reducing its wastage by 15%.

For academic studies, the first M.Sc. Course on operation research was started by Delhi University in 1963. At the same time, Institutes of Management at Calcutta and Ahmedabad introduced operation research in their MBA courses. At present, this subject has been introduced in almost all Institutes and Universities for the students of Mathematics, Statistics. Commerce, Economics, Management and Engineering

Some of Indian organisations using operation research techniques are: Indian Airlines, Railways, Defence organisations, Fertiliser Corporation of India, Hindustan Steel Ltd., Tata Iron and Steel Co., TELCO, DCM, CSIR. STC, etc. Assignment models have been used by Kirloskar company for allocation of their salesmen to different areas so us to maximize the profit. The company has also used linear programming models assemble various diesel engines at the lowest possible cost. D.C.M., Calico and Binnys' have been using linear programming techniques for cotton blending. Many organisations are making use of PERT/CPM techniques for effective management and control of their construction projects. A number of organisations are utilising operation research techniques for solving problems relating to staffing, production planning, blending, product mix, maintenance, inspection, advertising, capital budgeting, investment and the like.

THE NATURE, SIGNIFICANCE OF OPERATION RESEARCH

As the term implies, OR involves research on military operations. This indicates the approach as well as the area of its applications. The operations research approach is particularly useful in balancing conflicting objectives (goals or interests) where there are many alternative courses of action available to the decision-makers. In a theoretical sense, the optimum decision must be one that is best for the organization as a whole. It is often called the global optimum. A decision that is best for one or more sections of the organization is usually called suboptimum decision. The OR approach attempts to find global optimum by analysing interrelationships among the system components involved in the problem. One such situation is described below.

Consider a large organization with a number of management specialists but not necessarily well-coordinated. For example, consider the basic problem of maintaining stocks of furnished goods. To the marketing manager, stocks of a large variety of products are a means of supplying the company's customers with what they want and when they want it. Clearly, according to a marketing manager, a fully stocked warehouse is of prime importance to the company. But the production manager argues for long production runs preferably on a smaller product range; particularly if significant time is lost when production is switched from one variety to another. The result would again be a tendency to increase the amount of stock carried but it is, of course, vital that the plant should be kept running. On the other hand, the finance manager sees stocks in terms of capital tied up unproductively and argues strongly for their reduction. Finally, there appears the personnel manager for whom a steady level of production is advantageous for having better labour relations. Thus, all these people would claim to uphold the interests of the organization, but they do so only from their specialized points of view. They may come up with contradictory solutions and obviously, they cannot all be right.

In view of the above involving the whole system, the decision-maker, whatever his specialization, will need help and it is in the attempt to provide this assistance that OR has been developed. Operations research attempts to resolve the conflicts of interest among various sections of the organization and seeks the optimal solution which may not be acceptable to one department but is in the interest of the organization as a whole. Further, OR is concerned with providing the decision-maker with decision aids (or rules) derived from:

(i) A total system orientation,

(ii) Scientific methods of investigation, and

(iii) Models of reality, generally based on quantitative measurement and techniques.

Thus, successful application of OR techniques for solving a problem must involve the following steps:

1. Analysing the relationships among different variables and/or parameters associated with the problem, so as to determine consequences of decision alternatives.
2. Suggesting suitable measures of desirability (effectiveness or objective function) in order to evaluate the relative merit of decision alternatives (courses of action, acts or strategies).
3. Constructing mathematical, economic or statistical model of the problem under study to treat situations of complexity and uncertainty. This helps to view the problem in its entirety.

ADVANTAGES OF MODELS

Models in general are used as an aid for analysing complex problems. However, a model can also serve other purposes as given below:

1. A model allows us to analyse and experiment in a complex situation to a degree that would be impossible in the actual system and its environment. For example, the experimental firing of INSAT satellite may cost millions of rupees and require years of preparation.
2. Models simplify the investigation considerably and provide a powerful and flexible tool for predicting the future state of the process or system.

3. A model provides economy in representation of the realities of the system. That is, models help the decision-maker to visualize a system so that he can understand the system's structure or operation in a better way. For example, it is easier to represent a factory layout on paper than to construct it. It is cheaper to try out modifications of such systems by rearrangement on paper.
4. The problem can be viewed in its entirety, with all the components being considered simultaneously.
5. Models serve as aids to transmit ideas and visualization among people in the organization. For example, a process chart can help the management to communicate about better work methods to workers.

METHODOLOGY OF OPERATIONS RESEARCH

Every Operation research specialist may have his/her own way of solving problems. However, for effective use of Operation research techniques, it is essential to follow some steps that are helpful for decision-makers to make better decisions.

Step 1. Solving the Mathematical Model

Once a mathematical model of the problem has been formulated, the next step is to solve it, that is, to obtain numerical values of decision variables. Obtaining these values depends on the specific form or type, of mathematical model. Solving the model requires the use of various mathematical tools and numerical procedures. In general, the following two categories of methods are used for solving an OR model.

(i) *Optimization Methods:* These methods yield the best values for the decision variables both for unconstrained and constrained problems. In constrained problems, these values simultaneously satisfy all of the constraints and provide an optimal or acceptable value for the objective function or measure of effectiveness. The solution so obtained is called the optimal solution to the problem.

(ii) *Heuristic Methods:* These methods yield values of the variables that satisfy all the constraints, but not necessarily provide optimal solution. However, these values provide an acceptable value for the objective function.

Heuristic methods are sometimes described as rules of thumb which work. An example of a commonly used heuristic is 'stand in the shortest line'. Although using this rule may not work if everyone in the shortest line requires extra time, in general, it is not a bad rule to follow. These methods are used when obtaining optimal solution is either very time consuming or model is complex.

Step 2. Implementing the Solution

The decision-maker has not only to identify good decision alternatives but also to select alternatives that are capable of being implemented. It is important to ensure that any solution implemented is continuously reviewed and updated in the light of a changing environment. The behavioural aspects of change are exceedingly important to the successful implementation of results. In any case, the decision-maker who is in the best position to implement results must be aware of the objective, assumption, omissions and limitations of the model.

Step 3. Validating the Solution

After solving the mathematical model, it is important to review the solution carefully to see that the values make sense and that the resulting decisions can be implemented. Some of the reasons for validating the solution are:

(i) Certain aspects of the problem may have been overlooked, omitted or simplified.

(ii) The data may have been incorrectly estimated or recorded, perhaps when entered into the computer.

(iii) The mathematical model may not have enumerated all the limitations of the problem under consideration.

Step 4. Establishing Controls over the Solution

The dynamic environment and changes within the environment can have significant implications regarding the continuing validity of models and their solutions. Thus, a control procedure has to be established for detecting significant changes in decision variables of the problem so that suitable adjustments can be made in the solution without having to build a model every time a significant change occurs.

Step 5. Modifying the Model

For a mathematical model to be useful, the degree to which it actually represents the system or problem being modelled must be established. If during validation, the solution cannot be implemented, one needs to identify constraint that were omitted during the original problem formulation or to find if some of the original constraints were incorrect and need to be modified. In all such cases, one must return to the problem formulation step and carefully make the appropriate modifications to represent more accurately the given problem.

A model must be applicable for a reasonable time period and should be updated from time to time, taking into consideration the past, present and future aspects of the problem.

Step 6. Formulating the Problem

Problem formulation involves an analysis of the system under study, the objectives of the decision-maker, and alternative courses of action; etc., so as to understand and describe, in precise terms, the problem that an organization faces.

The analysis begins by detailed observation of the organizational structure, climate, communication and control system, its objectives and expectations. Such information will help in assessing the difficulty of the study in terms of costs, time requirement, resource requirement, probability of success of the study, etc.

The major steps which have to be taken into consideration for formulating the problems are as follows:

Decision Environment: It is desirable to know about the resources such as managers, employees, equipments, etc., which are required to carry out the policies of the organization considering the social and ecological environment in which the organization functions. Knowledge of such factors will help in modifying the initial set of the decision-maker's objectives.

Alternative courses of action The problem arises only when there are several courses of action available for a solution. An exhaustive list of courses of action can be prepared in the process of going through the above steps of formulating the problem. Courses of action which are not feasible with respect to objectives and resources may be ruled out.

Measure of Effectiveness: A certain measure of effectiveness or performance is required in order to evaluate the merit of the several courses of action listed above. The performance or effectiveness can be measured in different units such as rupees (net profits), percentage (share of market desired), time dimension (service or waiting time). In order to bring uniformity in the measurement, one of the performance criteria must be chosen to express the value in terms of the value of the criterion chosen earlier.

Problem Components: The first component of the problem to be defined is the decision-maker who is not satisfied with the existing state of affairs.

The interaction with the decision-maker will help the OR specialist in knowing his objectives. That is, either he has already obtained some solution of the problem and wants to retain it, or he wants to improve it to a higher degree. If the decision-maker has conflicting multiple objectives, he may be advised to rank his objectives in order of preference; overlapping among several objectives may be eliminated.

Qualitative objectives which cannot be quantified must be assigned weightage and probability measures to their attainability.

Step 7. Collecting Data and Constructing a Mathematical Model

After the problem is clearly defined and understood, the next step is to collect required data and then formulate a mathematical model. Model construction consists of hypothesizing relationships' between variables subject to and not subject to control by decision-maker. Certain basic components required in every decision problem model are:

Uncontrollable Variables: These are the factors whose numerical value depends upon the external environment prevailing in the organization. The values of these variables- are not under the control of the decision-maker and are also termed as state of nature.

Objective Function: It is a representation of (i) the criterion that expresses the decision-maker's manner of evaluating the desirability of alternative values of the decision variables, and (ii) how that criterion is to be optimized (minimized or maximized). For example, in queuing theory the decision-maker may consider several criteria such as minimizing the average waiting time of customers, or the average number of customers in the system at any time.

Controllable (Decision) Variables: These are the issues or factors in the problem whose values are to be determined (in the form of numerical values) by solving the model. The possible values assigned to these variables are called decision alternatives (strategies or courses of action). For example, in queuing theory, the number of service facilities is the decision variable.

Constraints (or Limitations): These are the restrictions on the values of the decision variables. These restrictions can arise due to limited resources such as space, money, manpower, material, etc. The constraints may be in the form of equations or inequalities.

Functional relationships In a decision problem, the decision variables in the objective function and in the constraints are connected by a specific functional relationship. A general decision problem model might take the form:

Optimize (Max or Min) $Z = f(x)$,

subject to the constraints

$$g_i(x) \{\leq, =, \geq\} b_i; \quad i = 1, 2, ..., m$$

and $\quad x \geq 0$

where, $\quad x$ = a vector of decision variables (xp x^, . . ., x^)

$f(x)$ = criterion or objective function to be optimized

$g_i(x)$ = the ith constraint

b_i = fixed amount of the ith resource.

A model is referred to as a linear model if all functional relationships among decision variables $x_1, x_2..., x$ in $f(x)$ and $g(x)$ are of a linear form. But if one or more of the relationships are non-linear, the model is said to be a non-linear model.

Parameters: These are constants in the functional relationships. Parameters can be deterministic or probabilistic in nature. A deterministic parameter is one whose value is assumed to occur with certainty. However, if constants are considered directly or explicitly as random variables, they are probabilistic parameters.

OPPORTUNITIES AND SHORTCOMINGS OF OPERATIONS RESEARCH APPROACH

The use of quantitative methods is appreciated to improve managerial decision-making. However, besides certain opportunities, OR approach has not been without its shortcomings. The main reasons for its failure are due to unawareness on the part of decision-makers about their own role, as well as avoiding the behavioural/organizational issues while constructing a decision model. A few opportunities and shortcomings of the OR approach are listed below.

Opportunities

1. It compels the decision-maker to be quite explicit about his objective, assumptions and his perspective to constraints.
2. It makes the decision-maker consider very carefully just what variables influence decisions.
3. Quickly points out gaps in the data required to support workable solutions to a problem.
4. Its models can be solved by a computer, thus the management can get enough time for decisions that require quantitative approach.

Shortcomings

1. Often solution to a problem is derived either by making it simplified or simplifying assumptions and thus, such solutions have limitations.
2. Sometimes models do not represent the realistic situations in which decisions must be made.
3. Often decision-maker is not fully aware of the limitations of the models that he is using.
4. Many 'real world' problems just cannot have an operation research solution.

FEATURES OF OPERATIONS RESEARCH SOLUTION

A solution that works but is quite expensive compared to the potential savings from its application should not be considered successful. Also a solution that is well within the budget but which does not accomplish the objective is not successful either. A few requirements of a good solution are as under.

1. *Technically Appropriate :* The solution should work technically, meet the constraints and operate in the problem environment.
2. *Reliable :* The solution must be useful for a reasonable period of time under the conditions for which it was designed.
3. *Economically Viable :* The economic value should be more than what it costs to develop and it should be seen as a wise investment in hiring Operation research talent.
4. *Behaviourally Appropriate :* The solution should be behaviourally appropriate and must remain valid for a reasonable period of time within the organization.

APPLICATIONS OF THE OPERATIONS RESEARCH

Some of the industrial/government/business problems which can be analysed by operation research approach have been arranged by functional areas as follows.

Finance and Accounting

- Dividend policies, investment and portfolio management, auditing, balance sheet and cash flow analysis.
- Break-even analysis, capital budgeting, cost allocation and control, and financial planning.

- Claim and complaint procedure, and public accounting.
- Establishing costs for by-products and developing standard costs.

Purchasing, Procurement and Exploration

- Optimal buying and reordering with or without price quantity discount
- Bidding policies
- Predicting customer loyalty
- Transportation planning
- Vendor analysis.

Marketing

- Selection of product-mix, marketing and export planning
- Best time to launch a new product
- Sales effort allocation and assignment
- Advertising, media planning, selection and effective packing alternatives
- Predicting customer loyalty.

Production Management

Facilities Planning

- Location and size of warehouse or new plant, distribution centres and retail outlets
- Logistics, layout and engineering design
- Transportation, planning and scheduling.

Manufacturing

- Aggregate production planning, assembly line, blending, purchasing and inventory control
- Employment, training, layoffs and quality control
- Allocating R&D budgets most effectively.

Maintenance and Project Scheduling

- Maintenance policies and preventive maintenance
- Maintenance crew size and scheduling
- Project scheduling and allocation of resources.

Techniques and General Management

- Decision support systems and MIS, forecasting
- Making quality control more effective
- Project management and strategic planning.

Government

- Economic planning, natural resources, social planning and energy
- Urban and housing problems
- Military, police, pollution control, etc.

Personnel Management

- Manpower planning, wage/salary administration
- Designing organization structures more effectively
- Negotiation in a bargaining situation
- Skills and wages balancing
- Scheduling of training programmes to maximize skill development and retention.

BASIC OPERATIONS RESEARCH MODELS

There is no unique set of problems which can be solved by using operation research models or techniques. Several operation research models or techniques can be grouped into some basic categories as given below. In this book, a large number of operation research models have been discussed in detail. Here, only an introductory description of these models is given.

Allocation Models

Allocation models are used to allocate resources to activities in such a way that some measure of effectiveness (objective function) is optimized. Mathematical programming is the broad term for the operation research techniques used to solve allocation problems.

If the measure of effectiveness such as profit, cost, etc., is represented as a linear function of several variables and if limitations on resources (constraints) can be expressed as a system of linear equalities or inequalities, the allocation problem is classified as a linear programming problem. But if the objective function of any or all of the constraints cannot be expressed as a system of linear equalities or inequalities, the allocation problem is classified as a non-linear programming problem.

When the solution values or decision variables for the problem are restricted to being integer values or just zero-one values, the problem is classified as an integer programming problem or a zero-one programming problem, respectively.

The problem having multiple, conflicting and incommensurable objective functions (goals) subject to linear constraints is called a Goal Programming problem. If the decision variables in the Linear Programming Problem depend on chance, such a problem is called a Stochastic Programming Problem.

If resources such as workers, machines or salesmen can be assigned to perform a certain number of activities such as jobs or territories on a one-to-one basis so as to minimize total time, cost or distance involved in performing a given activity, such problems are classified as assignment problems. But if the activities require more than one resource and conversely if the resources can be used for more than one activity, the allocation problem is classified as a transportation problem.

Dynamic Programming Models

Dynamic programming may be considered as an outgrowth of mathematical programming involving the optimization of multi-stage (sequence of inter-related decisions) decision processes. The method starts by dividing a given problem into stages or sub-problems and then solves those sub-problems sequentially until the solution to the original problem is obtained.

Waiting Line (or Queuing) Models

These models have been developed to establish a trade-off between costs of providing service and the waiting time of a customer in the queuing system. Constructing a model entails describing the components of the system: Arrival process, queue structure and service process and solving for the measure of performance—average length of waiting time, average time spent by the customer in the line, traffic intensity, etc., of the waiting system.

Competitive (Games Theory) Models

These models are used to characterize the behaviour of two or more opponents (called players) who compete for the achievement of conflicting goals. These models are classified according to several factors such as number of competitors, sum of loss and gain, and the type of strategy which would yield the best or the worst outcomes.

Replacement Models

These models are used when one must decide the optimal time to replace an equipment for one reason or the other—for instance, in the case of equipment whose efficiency deteriorates with time or fails immediately and completely. For example, in case of an automobile, the user has his own measure of effectiveness. So there will not be a single optimal answer for everyone even if each automobile gives exactly the same service.

Network Models

These models are applied to the management (planning, controlling and scheduling) of large-scale projects. PERT/CPM techniques help in identifying potential trouble spots in a project through the identification of the critical path. These techniques improve project coordination and enable the efficient use of resources. Network methods are also used to determine time-cost trade-off, resource allocation and updating of activity time.

Sequencing Models

The sequencing problem arises whenever there is a problem in determining the sequence (order) in which a number of tasks can be performed by a number of service facilities such as hospital, plant, etc., in such a way that some measure of performance, for example, total time to process all the jobs on all the machines, is optimized.

Decision Analysis Models

These models deal with the selection of an optimal course of action given the possible payoffs and their associated probabilities of occurrence. These models are broadly applied to problems involving decision-making under risk and uncertainty.

Markov-Chain Models

These models are used for analysing a system which changes over a period of time among various possible outcomes or states. The model while dealing with such systems describes transitions in terms of transition probabilities of various states. These models have been used to test brand-loyalty and brand-switching tendencies of consumers, where each system state is considered to be a particular brand purchase. These have also been used in reliability analysis, where the states of the system are the various levels of performance of the equipment being monitored.

Simulation Models

These models are used to develop a method to evaluate the merit of alternative courses of action by experimenting with a mathematical model of the problems where various variables are random. That is, these provide a means for generating representative samples of the measures of performance variables. Thus, repetition of the process by using the simulation model provides an indication of the merit of alternative course of action with respect to the decision variables.

Inventory Models

Inventory models deal with the problem of determination of how much to order at a point in time and when to place an order. The main objective is to minimize the sum of three conflicting inventory costs: The cost of holding or carrying extra inventory; the cost of shortage or delay in the delivery of items when it is needed, a cost of ordering or set-up. These are also useful in dealing with quantity discounts and selective inventory control.

ADVANTAGES OF OPERATIONS RESEARCH STUDY

A decision-maker in business and industry comes across the problem of making decisions in his day-to-day activities. In addition to learning about various operation research models useful in making a decision, the decision maker can also get several other advantages listed below.

Structured Approach to Problems

A substantial amount of time and effort can be saved in developing and solving operation research models if one uses a logical and consistent approach. This implies that the decision-maker has to be careful while defining controllable and uncontrollable variables, availability of resources—man, machine, material and money, etc., functional relationships among variables in the objective function and constraints. This will also reduce the chance of conceptual and computational errors. Any such error can also be detected easily and corrected.

Critical Approach to Problem Solving

The decision-maker will come to understand various components of the problem and accordingly select a mathematical model for solving the given problem. He will be aware of the explicit and implicit assumptions and inherent limitations of such models. Problem solutions are examined critically and the effect of any change and error in the problem data can be studied through sensitivity analysis techniques. In other words, the ability to judge applicability and limitations of solution techniques to a particular problem, as well as to critically analyse a proposed solution and data on which they are based, is bound to be a valued skill in any business organization.

DIFFICULTIES IN OPERATIONS RESEARCH

The previous sections have brought out the positive side of operation research only. However, there is also the need to point out the negative side. Certain common traps and pitfalls can and have, ruined the otherwise good work. Some of these pitfalls are quite obvious while others are so subtle and hidden that extreme care is required to locate their presence.

In the very *first* phase of operation research—the problem formulation phase—a number of pitfalls can and do arise. It is necessary that the right problem be selected and it must be completely and accurately defined. Is the right problem being solved? Is the scope considered wide and proper ? Will it result in optimization or only sub-optimization? Will the solution properly reflect the objectives as well as the imposed constraints? Are proper effectiveness measures being used? This phase of problem formulation is perhaps the most important and touchest part of operation research study.

Secondly, data collection may also consume a very large portion of lime and money spent on operation research study.

Thirdly, the whole study by operations analyst is based on his observations in the past. Strictly speaking, these observations can only be related to the laws that operated in the past, as there is no evidence that the laws will continue to operate in future also. If the laws are applied to the future, it clearly amounts to extrapolation in time.

Fourthly, the operations researcher, while making observations, may affect the behaviour of the system he is studying. Moreover, however comprehensive his experiments may be, his observations can never be more than a sample of the whole. These difficulties present special hazards to operations researcher. His aim is to find out what happens in a working organization. He can get the information in two ways : by direct observations or from the previous records. The behaviour of an organization depends upon the activities of the persons in it and the very fact that they are being observed is bound to affect their behaviour. On the other hand, accuracy of previous records is always doubtful and they seldom provide trie complete information in all the points sought.

Perhaps the greatest difficulty in operation research, however, is created by the time factor. The managers have to make decisions one way or the other, and a fairly good solution to the problem at the right time may be much more useful than the perfect solution too late. Further, the cost involved is also an important factor. Sometimes, some simple application of operation research may yield a good solution quickly and it may be unwise to spend a lot of money and effort to produce a slightly better solution much later.

Other pitfalls in problem solving include:

(i) Warping the problem to fit a standard model, tool or technique,

(ii) Failure to test the model and solution before implementation, and

(iii) Failure to establish proper controls.

It is the responsibility of operation research scientist to translate his highly specialised and technical thoughts, ideas and concepts into simple operational procedures capable of being easily understood by the management and workers alike He must also ascertain that the new proposals are properly implemented. But for the proper implementation, the whole operation research study becomes useless.

Lastly, what may appear to be a pitfall is the fact that operation research study may raise more questions than it answers. However, this may ultimately result in more deep insight into the system, yielding further benefits and improvements.

LIMITATIONS OF OPERATIONS

1. Mathematical models which are essence of operation research do not take into account qualitative factors or emotional factors which are quite real. All influencing factors which cannot be quantified find no place in mathematical models.
2. Mathematical models are applicable to only specific categories of problems.
3. Being a new field, generally there is a resistance from the employees to the new proposals.
4. Management, who has to implement the advised proposals, may itself offer a lot of resistance due to conventional thinking.
5. Young enthusiasts, overtaken by its advantages and exactness, generally forget that operation research is meant for men and not that men are meant for it.

Thus, at the implementation stage, the decision cannot be governed by quantitative considerations alone. It must take into account the delicacies of human relationships. That is in addition to being a pure scientist, one has to be tactful and learn the an of getting the decisions implemented. This art can be achieved by experience as well as by getting training in social sciences, particularly psychology.

In fact, many managers may make a joke of operation research as they think that the decisions made otherwise may be better. But being aware of its limitations, they need to be convinced of its utility, which doubtlessly forms the essential guideline for making better decisions.

EXERCISES

1. "Operation research is the application of scientific methods, techniques and tools to problems involving the operations of a system so as to provide those in control of the system with optimum solutions to the problems." Discuss.
2. State the different types of models used in operation research. Explain briefly the general methods for solving these operation research models.
3. Is operations research a discipline, a profession, a set of techniques, a philosophy", or a new name for an old thing? Comment.
4. What is operations research? Describe briefly its applications.
5. Write a critical essay on the definition and scope of operations research.
6. What post-World War-II factors were important in the development of Operations research?
7. Discuss the significance and scope of operations research in modern management.
8. What were the significant characteristics of operation research applications during World War-II? What caused the discipline of Operation research to take on these characteristics during that period?
9. Comment on the following statements:
   (a) Operation research is the art of winning war without actually fighting it.
   (b) Operation research is the art of finding bad answers where worse exist.
   (c) Operation research replaces management by personality.
10. Define operation research and discuss its scope.
11. Discuss the significance and scope of operations research in modern management.

12. Quantitative techniques complement the experience and judgement of an executive in decision-making. They do not and cannot replace it. Discuss.
13. Explain the differences between scientific management and operations research.
14. Operation research is not more than a quantitative analysis of the problem.
15. Operations research is an ongoing process. Explain this statement with examples.
16. Distinguish between model results that recommend a decision and model results that are descriptive.
17. 'Mathematics tends to lull the unsuspecting into believing that he who thinks elaborately think well'. Do you think that the best operation research models are the ones that are most elaborate and complex mathematically? Why?
18. Operation research replaces management by personality.
19. Decision-makers are quick to claim that quantitative analysis talks to them in a jargon that does not sound like English. List four terms that might not be understood by a manager. Then explain in non-technical terms what each term means.
20. It is said that operations research increases the creative capabilities of a decision-maker. Do you agree with this view ? Defend your point of view with examples.
21. What is the purpose of a mathematical model? How does a model achieve this purpose? Consider in your answer the concept that a model is an abstraction of reality.
22. Discuss in detail the three types of models with special emphasis on important logical properties and the relationship the three types bear to each other and to modelled entities.
23. 'Operations research is the application of scientific methods, techniques and tools to problems involving the operations of a system so as to provide those in control of the system with optimum solutions to the problem.' Discuss.
24. "What are the situations where operations research techniques will be applicable?"
25. Assume that you have to make a decision. Describe your decision with specific reference to:

(a) The alternative courses of action available.

(b) The criteria you have set for the evaluation of the available alternatives.

(c) The variables with which you propose to construct a model to assist you in making the decision.

(d) The extent to which the variables and the relationships between them are quantifiable.

26. Comment on the following statements:

    (a) Operation research is a bunch of mathematical techniques.

    (b) Operation research is no more than a quantitative analysis of the problem.

    (c) Operation research advocates a systems approach and is concerned with optimization. It provides a quantitative analysis for decision-making.

    (d) Operation research has been defined semi-facetiously as the application of big minds to small problems.

27. 'Operations research is an aid for the executive in making his decisions by providing him with the needed quantitative information, based on the scientific method analysis.' Discuss this statement in detail, illustrating it with operation research methods that you know.

28. Discuss the points to justify that the primary purpose of operations research is to resolve the conflicts resulting from the various subdivisions of the functional areas like production, marketing, finance and personnel in an optimal, near optimal or satisfying way.

29. Why is the study of operations research important to the decision-maker?

30. How far can quantitative techniques be applied in management decision-making? Discuss, in detail, with special reference to any functional area of management pointing out their limitations, if any.

31. In constructing any operation research model, it is essential to realize that the most important purpose of the modelling process is 'to help manager better'. Keeping this purpose in mind, state any four operation research models that can be of help to chartered accountants in advising their clients.

32. Explain the scope and methodology of operation research, the main phases of operation research and techniques used in solving an operation research problem.

33. What is the role of operations research in decision-making?

34. How can operation research models be classified? What is the best classification in terms of learning and understanding the fundamentals of operation research?

35. Describe briefly the applications of operation research in managerial decision-making. Give suitable examples from different areas.

36. Explain the concept, scope and tools of operation research as applicable to business and industry.

37. Operations research increases creative and judicious capabilities of a decision-maker. Comment.

38. Define an operation research model and give four examples. State their properties, advantages and limitations.

39. What are the advantages and disadvantages of operations research models? Why is it necessary to test models and how would you go about testing a model?

40. Briefly trace the history of operations research. What happened to the development of operations research after World War-II?

41. State any four areas for the application of operation research techniques in financial management. How do they improve the performance of an organization?

42. It is common in business to insure against the occurrence of events which are subject to varying degrees of uncertainty, for example, ill health of senior executives. At the same time, the use of formal analytical models to assist in the process of making decisions on business problems which are generally subject to uncertainty does not appear to be very widespread.

    Describe the model building approach to the analysis of business problems under conditions of uncertainty. Discuss the apparent inconsistency in a company's willingness to insure when formal analytical models of an operations research nature which allow for uncertainty are relatively employed.

43. What are the situations (areas) where operation research techniques will be applicable?

44. Discuss the following:

    (a) Operation research is an inter-disciplinary approach,

    (b) Scientific method in operation research.

(c) Operation research is more than a quantitative analysis of the problem.

45. What are the essential characteristics of operations research? Mention different phases in an operations research study. Point out its limitations, if any.

46. Is operations research a discipline or a profession or a set of techniques or a philosophy or a new name for old things? Discuss.

47. What is a model? Discuss various classification schemes of models.

    (a) Discuss, in brief, the role of operation research models in decision-making. What areas of operation research have made a significant impact on the decision-making process?

    (b) Discuss the importance of operations research in the decision-making process.

    (c) Give some examples of various types of models. What is a mathematical model? Develop two examples of mathematical models.

48. What reasons can you think of to explain the fact that many models are built and never implemented? Does the absence of implementation mean that the entire model development activity was a waste.

49. 'Quantifying the elements of a decision problem is the easy part; the hard part is solving the model.' Do you agree? Why or why not?

50. Define operations research as a decision-making science.

    (a) Give the main characteristics of operation research.

    (b) Discuss the scope of operation research.

51. (a) Outline the broad features of the judgement phase and the research phase of the scientific method in operation research. Discuss in detail any one of these phases.

    (b) What are various phases of the operation research problem? Explain them briefly.

52. What is the role of operations research in decision making?

53. Define operations research. Explain critically the limitations of various definitions as you understand them.

54. Give any three definitions of operations research and explain them.

55. Give reasons why most definitions of operations research are not satisfactory.

56. Model building is the essence of the operations research approach. Discuss.
57. What is meant by a mathematical model of a real situation? Discuss the importance of models in the solution of operation research problems.
58. (a) Discuss the various phases in solving an operation research problem.
    (b) State the phases of an operation research study and their importance in solving problems.
59. Discuss the role and scope of quantitative methods for scientific decision-making in a business environment.
60. Recent developments in the field of computer technology have enabled operations research to integrate its models into information systems and thus make itself a part of decision-making procedures of many organizations. Comment.
61. Explain various types of operation research models and indicate their applications to production, inventory and distribution systems.
62. Explain the distinction between
    (a) Static and dynamic models,
    (b) Analytical and simulation models.
    (c) Descriptive and prescriptive models.
63. Write briefly about the following:
    (a) Iconic models,
    (b) Analogue models.
    (c) Mathematical models (or symbolic models).
64. 'Whether in a private, non-profit, or public organization, the most important and the distinguishing function of a manager is problem solving. The field of quantitative methods is dedicated to aiding managers in their problem-solving efforts. This is accomplished through the use of mathematical models to analyse the problems.' Discuss.
65. 'Executives at all levels in business and industry come across the problems of making decisions at every stage in their day-to-day activities. Operations research provides them with various quantitative techniques for decision-making and enhances their ability to make long-range plans and solve everyday problems of running a business and industry with greater efficiency, competence and confidence.' Elaborate the statement with examples.

66. Comment on the following statements:
    (a) Operations research advocates a systems approach and is concerned with optimization.
    (b) Operations research replaces management by personality.
67. State the phases of operation Research study and their importance in solving problems.
68. Explain how and why operations research methods have been valuable in aiding executive decisions.
69. Operation research is an aid for the executive in making his decisions by providing him the needed quantitative information, based on scientific method analysis. Discuss.
70. Does the fact that operation Research takes the organizational point of view instead of the individual problem-centred point of view, generate constraints on its increased usage?
71. Suppose you are being interviewed by the manager of a commercial firm for a job in the research department which deals with the applications of quantitative techniques. Explain the scope and purpose of quantitative techniques and its usefulness to the firm. Give some examples of the applications of quantitative techniques in industry.
72. 'The hard problems are those for which models do not exist' Interpret this statement. Give some examples.
73. Briefly explain the uses of operation research techniques in India. How are they found useful by the business executives? Which of the three techniques are most commonly used in India? Why?
74. Explain the steps involved in the solution of an operations research problem.
75. Discuss the importance of operations research in the decision-making process.
76. Explain how and why operations research methods have been valuable in aiding executive decisions.
77. "Mathematics of operations research is mathematics of optimazation." Discuss.
78. Give some examples of various types of models. What is a mathematical model? Develop two examples of mathematical models.
79. Define an operation research model and give four examples. State their properties, advantages and limitations.

MULTIPLE CHOICE QUESTIONS

1. Explain the essential characteristics of the following types of models:

   (a) Allocation models (b) Competitive games

   (c) Inventory (d) Waiting line.

2. A physical model is example of

   (a) an iconic model (b) an analogue model

   (c) a verbal model (d) a mathematical model

3. An optimization model:

   (a) mathematically provides the best decision

   (b) provides decision within its limited context

   (c) helps in evaluating various alternatives constantly

   (d) all of the above.

4. Managerial decisions are based on

   (a) an evaluation of quantitative data

   (b) the use of qualitative factors

   (c) numbers produced by formal models

   (d) all of the above

5. The use of decision models

   (a) is possible when the variable's value is

   (b) reduces the scope of judgement and intuition known with certainty in decision-making

   (c) requires the knowledge of computer software use

   (d) none of the above

6. Operations research approach is

   (a) multi-disciplinary (b) scientific

   (c) intuitive (d) all of the above

7. Operations research analysts do not

   (a) predict future operations

   (b) build more than one model

   (c) collect relevant data

   (d) recommend decision and accept

8. For analysing a problem, decision-makers should normally study
   (a) its qualitative aspects (b) its quantitative aspects
   (c) both (a) and (b) (d) neither (a) nor (b)
9. Decision variables are:
   (a) controllable (b) uncontrollable
   (c) parameters (d) none of the above
10. A model is:
    (a) an essence of reality (b) an approximation
    (c) an idealization (d) all of the above
11. Every mathematical model
    (a) must be deterministic
    (b) requires computer aid for its solution
    (c) represents data in numerical form
    (d) all of the above.

2

Integer Programming

INTRODUCTION

Integer programming essentially deals with linear programs in which all or some of the variables in the optimal solution assume non-negative integer or discrete values. Such problems are called "*all integer*" (when all the variables are restricted to integer values) or "*mixed integer*" (when some of the variables are restricted to integer values). When all the decision variables are allowed to take values of either zero or one (like assignment problems), the problem is referred as zero-one programming problem. In linear programming, the variables are allowed to take any non-negative value within feasible region of the particular problem. The divisibility in Linear programming may not pose a serious problem in many practical situations, such as production of sugar, cement, steel, etc., but in certain situations, the divisibility assumption is unrealistic and completely unacceptable. In case of production of cars, trucks, chairs, selection of persons, etc., a solution with fraction is meaningless.

In linear programming, each of the decision variable as well as slack and/or surplus variable is allowed to take any real or fractional value. However, there are certain practical problems in which the fractional value of the decision variables has no significance. For example, it does not make sense saying 15 men working on a project or 16 machines in a workshop. The integer solution to a problem can, however, be obtained by rounding off the optimum value of the variables to the nearest integer value. This approach can be easy in terms of economy of effort, time and cost that might be required to derive an integer solution but this solution may not satisfy all the given constraints. Secondly, the value of the objective function so obtained may not be optimal value. All such difficulties can be avoided if the given problem, where an integer solution is required, is solved by integer programming techniques.

There are certain decision problems where decision variables make sense only if they have integer values in the solution. Capital budgeting, construction scheduling, plant location and size, routing and shipping schedule, batch size, capacity expansion, fixed charge, etc., are few problems which demonstrate the areas of application of integer programming.

Though, one approach to get integer solution to these problems, may be to round off the fractional values of the optimum values, it might result into a near to optimal solution or it may be required to change the availability of resources. Therefore, the task of rounding off the fractional values of the basic variables and at the same time satisfying the given set of constraints is not an easy job. These types of problems can conveniently be solved using the integer programming technique.

A class of optimization problems that are obtained from the general Linear Programming Model by imposing the additional restrictions of variables taking on only integer values, is called as an Integer Linear Programming problems. It is adaptable to situations that allow only integer numbers for the final solution. Although, various algorithms have been developed for Integer Linear Programming, none of these methods is totally reliable from the computational standpoint, particularly, when the number of integer variables is large. The computational difficulty, with available Integer Linear Programming algorithms, has led to use other means to "solve" the problem.

Further, in some cases, the deviation from the "exact" integer values, obtained by rounding off the fractional values, may be large enough to give an infeasible solution. Therefore, it needs to develop a systematic procedure to identify the optimal integer solution to such problems.

First time in 1956, R.R. Gomory suggested the systematic method to obtain an integer solution to an "all integer programming problem", and later on the method was extended to deal with the more complicated problems of "Mixed Integer Programming".

The areas of application of integer programming include allocation of salesmen to sales regions, capital budgeting, research and development, optimum location of warehouses, assignment problems, etc.

TYPES OF INTEGER PROGRAMMING PROBLEMS

Linear integer programming problems can be classified into three categories:

(i) *Pure (all) integer programming* problems in which all decision variables are required to have integer values.

(ii) *Mixed integer programming problems* in which some, but not all, of the decision variables are required to have integer values.

(iii) *Zero-one integer programming problems* in which all decision variables must have integer values of 0 or 1.

Various methods for solving an integer programming problem are summarized on the next page.

Now, we shall discuss two methods:

(i) Gomory's cutting plane method, and

(ii) Branch and Bound method for solving integer programming problems.

The pure integer linear programming problem in its standard form can be stated as follows:

Maximize $Z = c_1 x_1 + c_2 x_2 + ... + c_n x_n$

subject to the constraints

$$a_{11}x_1 + a_{12}x_2 + a_{1n}x_n = b_1$$
$$a_{21}x_1 + a_{22}x_2 + a_{2n}x_n = b_2$$
$$\vdots \qquad \vdots$$
$$a_{m1}x_1 + a_{m2}\, x_2 + ... + a_{mn}x_n = b_m$$

and $\quad a_1, x_2, ..., x_n \leq 0$ and are integers.

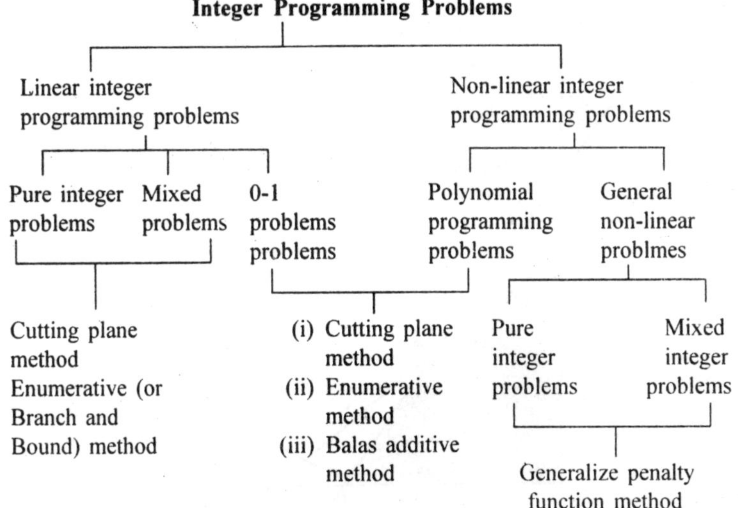

ENUMERATION AND CUTTING PLANE SOLUTION CONCEPT

Gomory's cutting-plane method was developed by R.E. Gomory in 1956 to solve integer linear programming problems using the dual simplex method.

It is based on the generation of a *sequence of linear inequalities called a cut*. This 'cut' cuts out a part of the feasible region of the corresponding LP problem while leaving out the feasible region of the integer linear programming problem. The hyperplane boundary of a cut is called the cutting plane.

Consider the following linear integer programming (LIP) problem

Maximize $Z = 14x_1 + 16x_2$

subject to the constraints

$$4x_1 + 3x_2 \le 12$$

$$6x_1 + 8x_2 \le 24$$

and $\quad x_1, x_2 \ge 2$ and are integers.

Relaxing the integer requirements, the problem is solved graphically using Fig. 2.1. We obtain the optimal solution to this LP problem as:

$$x_1 = 1.71,\ x_2 = 1.71$$

and $\quad \text{Max } Z = 51.42.$

This solution does not satisfy the integer requirement of variables x_1 and x_2. The feasible region (solution space) formed by the constraints is marked by OABC in Fig. 2.1.

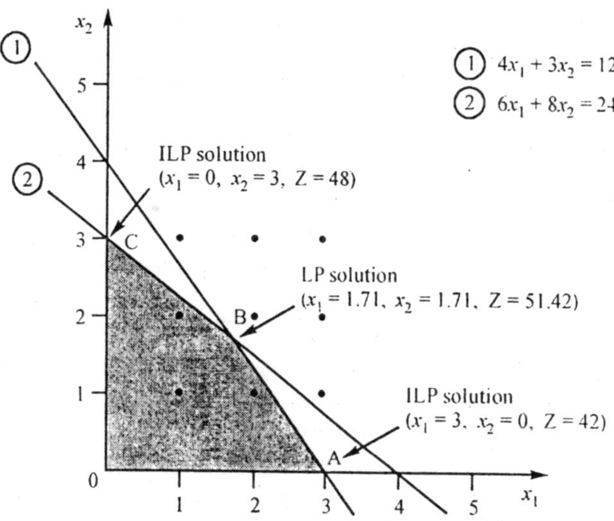

Fig. 2.1 : Concept of Cutting Plane.

Rounding off this solution to $x_1 = 2$, $x_2 = 2$ does not satisfy both the constraints and therefore, the solution is infeasible. The dots in Fig. 2.1, also referred to as *lattice points*, represent all of the integer solutions that lie within the feasible solution space of LP problem. However, it is difficult to evaluate every such point to determine the value of the objective function.

Figure 2.1 suggests that we can find a solution to the problem when it is formulated as an LP problem (which by chance could contain integers). It may be noted that the optimal lattice point C lies at the comer of the solution space OABC obtained by cutting away the small portion above the dotted line. This suggests a solution procedure that successively cuts down (reduces) the feasible solution space until an integer-valued corner is found.

The optimal integer solution to the given LP problem is: $x_1 = 0$, $x_2 = 3$ and Max $Z = 48$. Notice that its lattice point is not even adjacent to the most desirable LP problem solution comer.

Remark Reducing the feasible region by adding extra constraints (cut) can never give an improved objective function value. Usually it makes it worse and if Z_{IP} represents the minimum value of objective function in an ILP problem and Z_{LP} the minimum value of objective function in an LP problem, then $Z_{IP} \geq Z_{LP}$.

GRAPHICAL METHOD OF SOLVING INTEGER PROBLEM

Any IPP involving only two decision variables (x_1, x_2) can easily be solved by graphical method as applicable to IPP. The difference from LPP is that the optimal solution derived must have integer values. This can be explained with the help of the following illustration:

FORMULATION OF INTEGER PROGRAMMING PROBLEM

The integer programming problem requires the following characteristics to be satisfied:

(i) A linear objective function

(ii) A set of linear resource constraints

(iii) Integer value constraints for all or some variables, and

(iv) Non-negativity constraints for the decision variables.

Using the mathematical notations as used in formulating a Linear Programming Problem, Integer Programming Problem can be formulated mathematically as

$$\text{Optimize } \sum_{j=1}^{n} C_j x_j \quad \ldots(1)$$

Subject to $\sum_{j=1}^{n} a_{ij} x_j \le b_{ij}$ for i = 1, 2,......,m ...(2)

$x_j \ge 0$ for j = 1, 2...........n ...(3)

x_j are integer values for j = 1, 2,.......s ...(4)

Therefore, the Integer Programming Problem may be considered as a special case of Linear Programming with an additional constraints of type (4) included in the Linear Programming Problem. In (4) when s = n (every variable is integer values), the problem is referred as pure Integer Programming Problem, otherwise it is called a *Mixed Integer Programming Problem.* Depending upon the areas of application, the objective function (1) may be either maximization or minimization. An Integer programming problem may include ≥ in-equalities and/or equalities as well.

One of the easiest ways of solving the IPP is to transform the problem into a form, which allows the application of LP. Business situations of mixed IPP are quite common, *e.g.,* a metal container manufacturing firm should cut various sizes of blanks from the metal sheet stock with minimum waste. One of the restrictions is that the whole blanks should be cut from sheets. It is simple to round off non-integer solutions in problems involving indivisible resources, especially if the rounding off is small relative to the values of the variables involved. However, such rounding off can result in solutions far from the optimal integer solution. Following example demonstrates the Integer Programming Problem case:

Although, several methods are available for solving the all-integer and mixed IPP, Gomory's cutting plane method and branch-and-bound algorithms are more popular.

GOMORY'S ALL INTEGER CUTTING PLANE METHOD

Historically, the first method for solving Integer Programming Problems was the cutting plane method. There are different approaches to cuts, but in this text, R.E. Gomory's fractional cut will be discussed, since it guarantees an optimal solution in a finite number of iterations.

A systematic procedure called Gomory's all integer algorithm will be discussed here for generating 'cuts' (additional linear constraints) so as to ensure an integer solution to the given LP problem in a finite number of steps. Gomory's algorithm has the following properties:

(i) *Additional linear constraints* never cut-off that portion of the original feasible solution space which contains a feasible integer solution to the original problem.

(ii) *Each new additional constraint* (or hyperplane) cuts-off the current non-integer optimal solution to the linear programming problem.

Method for Constructing Additional Constraint (Cut)

Gomory's method begins by solving the LP problem without taking into consideration the integer value requirement of the decision variables. If the solution so obtained is an integer, *i.e.,* all variables in the 'x_B' *column* (also called basis) of the simplex table assume non-negative integer values, the current solution is the optimal solution to the given ILP problem. But, if some of the basic variables do not have non-negative integer value, an additional linear constraint called the Gomory constraint (or cut) is generated. After having generated a linear constraint (or cutting plane), it is added to the bottom of the optimal simplex table so that the solution no longer remains feasible. The new problem is then solved by using the dual simplex method. If the optimal solution so obtained is again non-integer, another cutting plane is generated. The procedure is repeated until all basic variables assume non-negative integer values.

The procedure of developing a cut is discussed below. In the optimal simplex table, we select one of the rows, called *source row* for which basic variable is non-integer. The desired cut is developed by considering only fractional parts of the coefficients in source row. For this reason, such a cut is also referred to as *fractional cut.*

Suppose the basic variable x_r has the largest fractional value among all basic variables restricted to be integers. Then the rth constraint equation (row) from the simplex table can be rewritten as:

$$X_{Br} (= B_r) = 1.x_r + (a_{r1}x_1 + a_{r2}x_2 + \ldots)$$

$$= x_r + \sum_{j \neq r} a_{rj}x_j \qquad \ldots(1)$$

where x_j (j = 1,2, 3,...) represents all the non-basic variables in the rth constraint (row) except the variables x_r and b_r (= x_{Br}) is the non-integer value of variable x_r. Let us decompose the coefficients of x_j, x_r. variables and x_{Br} into integer and non-negative fractional parts in Eqn. (1) as shown below:

$$[x_{Br}] + f_r = (1 + 0)\, x_r + \sum_{j \neq r} \left\{[a_{rj}] + f_{rj}\right\} x_j \qquad \ldots(2)$$

where $[x_{Br}]$ and $[a_{rj}]$ denote the largest integer obtained by truncating the fractional part from x_{Br} and a_{rj} respectively.

Rearranging Eqn. (2) so that all the integer coefficients appear on the left-hand side, we get

$$f_r + \{[x_{Br}] - x_r - \sum_{j \neq r} [a_{rj}] x_j\} = \sum_{j \neq r} f_{rj} x_j \qquad ...(3)$$

where f_r is a strictly positive fraction ($0 < f_r < 1$) while f_{rj} is a non-negative fraction ($0 \leq f_{rj} \leq 1$).

Since all the variables (including slacks) are required to assume integer values, terms in the bracket on the left hand side as well as on the right hand side must be non-negative numbers. Since the left-hand side in Eqn. (3) is f_r plus a non-negative number, we may write it in the form of the following inequalities:

$$f_r \leq \sum_{j \neq r} f_{rj} x_j \qquad ...(4)$$

or

$$\sum_{j \neq r} f_{rj} x_j = f_r + s_g \text{ or } -f_r = s_g - \sum_{j \neq r} f_{rj} x_j \qquad ...(5)$$

where s_g is a non-negative slack variable and is called the *Gomory Slack Variable.*

Equation (5) represents Gomory's cutting plane constraint. When this new constraint is added to the bottom of optimal simplex table, it would create an additional row to the table along with a column for the nsw variable s_g.

Steps of Gomory's All Integer Programming Algorithm

An iterative procedure for the solution of an all integer programming problem by Gomory's cutting plane method may be summarized in the following steps.

Step 1 : Initialization formulate the standard integer LP problem. If there are any non-integer coefficients in the constraint equations, convert them into integer coefficients. Solve it by simplex method, ignoring the integer requirement of variables.

Step 2 : Test the optimality (a) Examine the optimal solution. If all basic variables (*i.e.,* $x_{Br} = b_i \geq 0$) have integer values, the integer optimal solution has been derived and the procedure should be terminated. The current optimal solution obtained in Step 1 is the optimal basic feasible solution to the integer linear programming.

(b) If one or more basic variables with integer requirements have non-integer solution values, then go to Step 3.

Step 3 : Generate cutting plane Choose a row r corresponding to a variable x_r. which has the largest fractional value f_r and generate the cutting plane (a Gomory constraint) as explained earlier in eqn. (5);

$$-f_r = s_g - \sum_{j \neq r} f_{rj} x_j$$

where $0 \leq f_{rj} < 1$ and $0 < f_r < 1$

If there are more than one variables with the same largest fraction, then choose the one that has the smallest contribution to the maximization LP problem or the largest cost to the minimization LP problem.

Step 4 : Obtain the new solution Add the cutting plane generated in Step 3 to the bottom of the optimal simplex table as obtained in Step 3. Find a new optimal solution by using the dual simplex method, *i.e.*, choose a variable to enter into the new solution having the smallest ratio: $\{(c_j - z_j)/y_{ij}; y_{ij} < 0\}$ and return to Step 2. The process is repeated until all basic variables with integer requirements assume non-negative integer values.

The procedure for solving an ILP problem can be explained through a flow chart shown in Fig. 2.2.

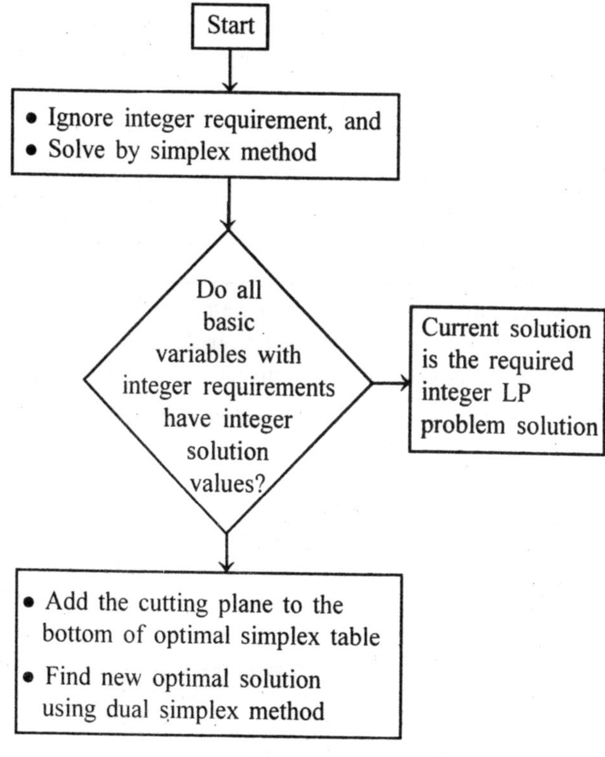

Fig. 7.2 : Flow Chart for solving ILP problem.

CONSTRUCTION OF GOMORY'S CUTTING PLANE CONSTRAINT

First the relaxed Linear Programming Problem is solved disregarding the integer condition. If it leads to optimal solution to be integer, no more computation would be required. If not, the secondary constraints forcing the solution toward the integer solution, are generated as follows:

Let the final optimal tabeau for Linear Programming be given as in Table 2.1.

Table 2.1

			Non basic					
Basic	x_1	$x_2 \ldots$	$x_i \ldots$	$x_m \ldots$	$w_i \ldots$	$w_j \ldots$	w_n	*Solution*
x_1	1	0	0	0	α_1^1	α_1^j	α_1^n	β_1
x_2	0	1	0	0	α_2^1	α_2^j	α_2^n	β_2
......								
......								
x_i	0	0	1	0	α_i^1	α_i^j	α_i^n	β_i
......								
......								
x_m	0	0	0	1	α_m^i	α_m^j	α_m^n	β_m
$\overline{Z}$	0	0	0	0	$\bar{c}_i$	$\bar{c}_j$	$\bar{c}_n$	

The variables x_i (1, 2,.......m) represent the basic variables and the variables w_j (j = 1, 2, ...n) are the non-basic variables. These variables have been arranged as such for convenience.

Now, if the *ith* equation is considered where basic variables x_i assumes a non-integer value,

$$x_i = x_i = \beta_i - \sum_{j=1}^{n} \alpha_i^j w_j. \quad [\beta_i \text{ non-integer (source row)}]$$

Any such equation will be referred as a source row. Since, in general, the coefficient of the objective function can be made integer, the variable $\overline{Z}$ is also integer and *Z* equation maybe chosen as a source row. In fact, the convergence proof of algorithm requires that $\overline{Z}$ be integer.

Let $\quad \beta_i = (\beta_i) + f_i$

$$\alpha_i^j = (\alpha_i^j) + f_{ij}$$

If N = (a) is the largest integer such that $N \le a$, it follows that $0 < f_i < 1$ and $0 \le f_{ij} < 1$, *i.e.*, f_i is strictly positive fraction and f_{ij} is a non-negative fraction. *e.g.*,

a	*(a)*	*f = a – (a)*
2½	2	1/2
–2½	–3	1/2
–2	–2	0
–3/5	–1	2/5

The source row thus yields

$$f_i - \sum_{j=1}^{n} f_{ij}\, w_j = x_i - (\beta_i) + \sum_{j=1}^{n} (\alpha_i^j) w_j$$

For all the variables x_i and w_j to be integer, the right hand side of the above equation should be integer, *i.e.*, the left hand side must also be integer.

Given $f_{ij} \ge 0$ and $w_j \ge 0$ for all *i* and *j*.

It follows that $\sum_{j=1}^{n} f_{ij}\, w_j \ge 0$

Consequently

$$f_i - \sum_{j=1}^{n} f_{ij}\, w_j \le f_i < 1$$

Since $f_i - \sum_{j=1}^{n} f_{ij}\, w_j$ must be integer by construction, an essential condition for satisfying the integerality becomes

$$f_i - \sum_{j=1}^{n} f_{ij}\, w_j \le 0$$

The last constraint can be added in the form

$$S_i = \sum_{j=1}^{n} f_{ij}\, w_j - f_i \text{ (fractional cut)}$$

Where S_i is a non-negative slack variable, which should be an integer (by definition). This constraint equation defines *fractional cut.* From the Table 2.1, $w_i = 0$, and thus $S_i = -f_i$, which is not feasible. This means that the new constraint is not satisfied by the given solution. The dual simplex method can then be used to clear the infeasibility, which is equivalent to cutting-off the solution space toward the optimal integer solution.

The new tableau after addition of the fractional cut, thus, becomes as given in Table 2.2.

Table 2.2

Basic	x_1	$x_2 \ldots$	$x_i \ldots$	$x_m \ldots$	$w_1 \ldots$ (*Non basic*)	$w_j \ldots$	$w_n \ldots$	S_i	*Solution*
x_1	1	0	0	0	α^1_1	α^j_1	α^n_1	0	β_1
x_2	0	1	0	0	α^1_2	α^j_2	α^n_2	0	β_2
…	…	…	…	…	…	…	…	…	…
…	…	…	…	…	…	…	…	…	…
x_i	0	0	1	0	α^1_i	α^j_i	α^n_i	0	β_i
…	…	…	…	…	…	…	…	…	…
…	…	…	…	…	…	…	…	…	…
x_m	0	0	0	1	α^i_m	α^j_m	α^n_m	0	β_m
S_i	0	0	0	0	$-f_{i1}$	$-f_{ij}$	$-f_{in}$	1	$-f_i$
Z	0	0	0	0	$\bar{c}_1$	$\bar{c}_j$	$\bar{c}_n$	0	

If the solution (after applying dual simplex method) is integer, the computation ends. If not, a new fractional cut is generated from the *resulting tableau* and the dual simplex method is again used to clear the infeasibility. This procedure is repeated until an integer solution is achieved. However, if any iteration in the dual simplex algorithm indicates that no feasible solution exists, the problem has no feasible *integer* solution.

The algorithm is called as the *fractional method* because all the non-zero coefficients of the generated cuts are less than one.

At the first instance, the fractional algorithm may indicate that the size of the simplex tableau may become very large with new cuts included in the problem. This is not true and infact, the total number of constraints in the augmented problem cannot exceed the number of variables in the original problem, *i.e.*, (m + n). This follows that if the augmented problem includes more than (m + n) constraints, one or more of the slack variables S_i associated with the fractional cuts must become basic. In such a case, the associated equations become redundant and may be dropped from the tableau completely.

The fractional algorithms have two disadvantages:

1. The round-off errors involved in automatic calculations will most likely distort the original data, particularly with the increase in problem size.

2. The solution of the problem remains infeasible in the sense that no integer solution can be obtained until the optimal integer solution is reached. This means that there will be no "good" integer solution in store, if the calculations are stopped prematurely prior to the attainment of the optimal (integer) solution.

The first difficulty is overcome by the development of an *all-integer algorithm,* which starts with an initial all-integer tableau (*i.e.,* all the coefficients are integers) suitable for the application of the dual simplex algorithm. Special cuts are then generated such that their addition to the tableau will not lose the integerability of all the coefficients. However, since the solution still remains infeasible until the integer optimal solution is arrived, still presents a disadvantage.

The second difficulty was considered by developing cutting-plane algorithms that start integer and feasible but non-optimal. The iterations continue to be feasible and integer until the optimum solution is reached. In this respect, this algorithm is *primal feasible* as compared with the fractional algorithm, which is dual-feasible. The primal algorithms do not appear to be computationally promising, however.

RULES FOR CUT (STRENGTH OF THE FRACTIONAL CUT)

The specific inequality defining a cut depends directly on the "source row" from which it is developed. Thus, different inequality cuts may be developed from the same simplex tableau. The obvious question is, which cut is the "strongest"? Strength could be measured in terms of how deep the inequality cuts into the solution space. This can be expressed mathematically as below:

Consider the two inequalities

$$\sum_{j=1}^{n} f_{ij}\, w_j \geq f_i \qquad (1)$$

and

$$\sum_{j=1}^{n} f_{kj}\, w_j \geq f_k \qquad (2)$$

Cut (1) is said to be stronger than (2) if $f_i \geq f_k$ and $f_{ij} \geq f_{kj}$ with the strict inequality holding at least once.

This definition of strength is difficult to implement computationally. Thus, emperical rules reflecting this definition are derived. Two such rules call for developing the cut from the source row that has (1) $\max_i \{f_i\}$ or

(2) $\max_i \left\{ f_i \sum_{j=1}^{n} f_{ij} \right\}$. The second rule is more effective, since it represents more closely the definition of strength given.

The optimum continuous solution is Z = 63, $x_1 = 7/2$, $x_2 = 9/2$. Since Z is already integer, its equation cannot be taken as source row. According to the emperical rules given, since $f_1 = f_2 = 1/2$, the rule is non-conclusive about which source row may be better. But to apply the second rule, it is necessary to develop all the coefficients of the respective fractional cuts from each source row. The cuts from the x_1 row and x_2 row are

$$x_1 \text{ row}: \quad \frac{7}{22}x_3 + \frac{1}{22}x_4 \geq \frac{1}{2}$$

$$x_2 \text{ row}: \quad \frac{21}{22}x_3 + \frac{3}{22}x_4 \geq \frac{1}{2}$$

$$\text{Since} \quad \frac{1/2}{7/22 + 1/22} > \frac{1/2}{21/22 + 3/22}$$

The x_1-equation is selected as source row. The cut from the x_1 and x_2 is given by

$x_2 \leq 3$.

By following a similar substitution, the cut from x_2-row is expressed as $x_1 \leq 10/3$

The first cut is more restrictive and hence stronger than the second cut.

BRANCH AND BOUND METHOD

Branch and Bound Method is a generic technique for solving a class of optimization problems. The Branch and Bound (B and B) method was originally developed by A.K. Land and A.G. Doig to solve the all integer, mixed integer and zero-one programming problem. The optimal solution is obtained by successive partitioning of the set of all possible solutions into two mutually exclusive and exhaustive subsets, and establishing limit for each set such that all solutions in the set are worse than the limit. The first part of the technique involving partitioning is known as *branching*, whereas the second part of establishing limit is called as *bounding*. Thus, in this method also, first a continuous integral programming problem is solved ignoring the integer valued condition. Branching signifies partitioning a current solution space into mutually exclusive subspaces. In the optimal solution, if some one of the variables, say x_j is not an integer, given integral programming problem is divided (partitioned) into two sub-problems. If x^*_j is the value of x_j in the optimal solution, then the following should be noted:

$$[x^*_j] < x_j < [x^*_j] + 1$$

Where $[x^*_j]$ is the integer part of the value of x_j.

Therefore, any feasible integer value of x_j should satisfy one of the following two conditions:

$$x_j \leq [x_j] \text{ and } x_j \geq [x_j] + 1$$

It may be noted that the above two constraints are mutually exclusive, *i.e.*, both cannot be true simultaneously and, therefore, both cannot be considered simultaneously in the integer programming problem. x_j is referred to as the *branching variable.*

Now two sub-integer programming problems are formed by adding these constraints separately to the continuous integer programming problem. Thus, an original problem is branched into two sub-problems. The branching process discards that portion of feasible region which involves no feasible integer solution. For example, in an optimal solution without the integer constraint, if $x_1 = 5/2$.

This obviously gives that $2 < x^*_1 < 3$, *i.e.*, in an integer valued solution, either $x_1 \leq 2$ or $x_1 \geq 3$. Thus, there will be no integer valued feasible solution in the space between $x_1 = 2$ and $x_1 = 3$.

The basics of branch and bounding algorithm can be more conveniently explained with the help of a numeric example.

AREAS OF APPLICATIONS OF INTEGER PROGRAMMING

The integer programming is considered as the technique to solve problems in linear programming structure involving the variables non-divisible. In practical business problems, there are many such situations, such as *capital budgeting, assignment problem, travel salesmen problems*, etc. Few such examples are discussed below:

Capital Budgeting Problems

This involves the situation of alternative decisions of investing or not investing. A mathematical model for capital budgeting is given by Weingartner.

If b_j = Net present value of *j*th investment proposal

x_j = An amount between 0 and 1 such that $x_j = 0$ when investment is not accepted and $x_j = 1$ when the investment is accepted.

C_{ij} = Net cash outflow required for the *j*th proposal in the *period i.*

B_i = Budget availability during the *i*th period.

Maximize $\sum_{j=1}^{n} b_j x_j$

Subject to $\sum_{j=1}^{n} c_{ij} x_j \leq B_i$ $\quad i = 1, 2, \ldots\ldots\ldots m.$

$x_j = 0$ or 1

The model maximizes a net value from currently available investment proposals with given budget constraints in each period.

Fixed Cost (or Charge) Problems

In certain cases, while undertaking a particular set of activities, the fixed costs (fixed charge or setup costs) are incurred. In such cases, the objective is the minimization of the total cost (sum of fixed and variable costs) associated with an activity:

Let us define the following decision variables

x_j = level of activity j

F_j = fixed cost associated with activity $x_j > 0$

c = variable cost associated with activity $x_j > 0$

Then the general fixed cost problem can be stated as:

$$\text{Minimize } Z = \sum_{j=1}^{n} (c_j x_j + F_j y_j)$$

subject to the constraints

$$\sum_{j=1}^{n} a_{ij} x_j \leq b_j; \qquad i = 1, 2, \ldots., m$$

$x_j \leq My_j$

or $\quad x_j - My_j \leq 0; \; j = 1, 2, \ldots, n$

$x_j \geq$ for all j

and $\quad y_j = 0$ or 1 for all j

where the symbol M denotes a large number so that $x_j \leq M$.

Travelling Salesmen Problem

Suppose there are *n* cities with known distances between any two cities. A salesman has to start from his home city and visit each city once and then returns back to his starting point (say city-1). The objective is to minimize the total distance covered (or cost or time). The problem can be formulated as zero-one integer programming problem. The *Travelling Salesman Problem* can be formulated as following:

$$\text{Minimize } \sum_i \sum_j \sum_k d_{ij}\, x_{ijk}, \; i \neq j$$

where d_{ij} = distance from city *i* to city *j*, and

1, if the salesman travels from city *i* to city *j* and city *j* is the *k*th city travelled

0, otherwise

(*i, j, k* arc integers varying from 1 to *n*)

The constraints involved would be as following:

(i) $\sum_i \sum_j x_{ijk} = 1, \quad k = 1, 2 \ldots\ldots n$

$i \neq j$

i.e., only one directed are may be assigned to a specific value of *k*.

(ii) $\sum_i \sum_k x_{ijk} = 1, \quad i = 1, 2, \ldots\ldots n$

i.e., only other city may be reached from a specific city *i*.

(iii) $\sum_i \sum_k x_{ijk} = 1, \quad j = 1, 2, \ldots\ldots n$

i.e., only one other city can initiate a directed arc to a specified city j.

(iv) $\sum_i x_{ijk} = \sum_r x_{jr(k+1)}$ for all j and k

$i \neq j, \quad r \neq j$

This constraint ensures that the complete travel will consist of a connected arc. It is given that the kth directed arc ends at some specific city j, the (k + 1)th directed arc must start at the same city j.

Assignment Problem

It refers to the situation where it is required to produce atleast N units (sizes) of a certain product on n different machines (or by n skilled workers), who are capable of producing any of the parts. However, each machine/worker requires different amount of time in producing a given product.

If

P_j = Number of products ordered of type j

x_{ij} = Number of products produced by machine/worker i

t_{ij} = Amount of time required for machine/worker *i* to produce a product of type *j*.

Since the objective of the company is to minimize the amount of time taken to produce all the required products, the problem can be formulated as below:

$$\text{Minimize} \quad Z = \sum_{i=1}^{n} \sum_{j=1}^{N} t_{ij} x_{ij}$$

$$\text{Subject to} \quad \sum_{i=1}^{n} x_{ij} \geq P_j, \; j = 1, 2, \ldots N$$

x_{ij} are non-negative integers for all i and j.

If the cost aspect is introduced in the objective function, the integer programming problem could be somewhat modified. If each skilled worker/machine is allowed a total of h_i hours to complete his work (*e.g.*, $h_i = 8$, for $i = 1, 2, \ldots m$) and that the cost of having worker/machine i to produce a product of type j is C_{ij}, the problem then becomes

$$\text{Minimize} \quad Z = \sum_{i=1}^{n} \sum_{j=1}^{N} C_{ij} x_{ij}$$

$$\text{Subject to} \quad \sum_{i=1}^{n} x_{ij} \geq P_j, \; j = 1, 2, \ldots\ldots\ldots N$$

x_{ij} are non-negative integers for all i and j.

Plant Location Problem

Suppose there are m possible sites at which plants could be located. Each of these plants produces a single commodity for n customers each with a minimum demand required for b_j units ($j = 1, 2, \ldots, n$). The fixed set-up cost (expenses associated with constructing and operating a plant) for a plant in the rth location is f_i ($i = 1, 2, \ldots, m$). The production capacity for each plant is limited to a_i units. The unit transportation cost from plant j to customer j is c_{ij}. The problem is to locate the plants in such a way that the sum of the fixed set-up costs and transportation cost is minimized.

Let x_{ij} be the amount shipped from plant i to customer j, and y_i be the new variable associated with each of the possible plant locations, such that

$$y_i = \begin{cases} 1, \text{ if plant is located at the ith location} \\ 0, \text{ otherwise} \end{cases}$$

The value of f_i is assumed to be fixed and independent of the amount x_{ij} shipped so long as $x_{ij} > 0$, *i.e.*, for $x_{ij} = 0$, the valuer = 0, Now we can formulate the problem as zero-one integer programming problem as follows:

$$\text{Minimize } Z = \sum_{i=1}^{m} c_{ij}x_{ij} \sum_{j=1}^{n} c_{ij}x_{ij} + \sum_{i=1}^{m} f_i y_i$$

subject to the constraints

$$\sum_{i=1}^{m} x_{ij} \geq b_j; \qquad j = 1, 2, ..., n \qquad ...(i)$$

$$\sum_{j=1}^{n} x_{ij} \leq y_i u_i; \qquad i = 1, 2, ..., m \qquad ...(ii)$$

$$\sum_{j=1}^{n} x_{ij} \leq a_i; \qquad i = 1, 2, ..., m \qquad ...(ii)$$

and $\qquad x_{ij} \geq 0$, for all i and j.

Constraints ; (i) guarantee that each customer's demand is met. If all the shipping costs are positive, *i.e.,* there are no subsidized routes, then it never pays to send more than needed, or we replace the inequality sign by an equality. Inequality (ii) ensure that we do not ship from a plant which is not operating, u_i is the upper bound on the amount shipped which may be shipped from plant i and inequality (iii) restricts production from exceeding the limited capacity.

GOMORY'S MIXED-INTEGER CUTTING PLANE METHOD

In the previous section, it was important to note that the fractional cut assumes that all the variables, including slack and surplus, are integers. This means that the application of the fractional cut will yield no feasible integer solution unless all variables assume integer values. Simultaneously, for example, consider the constraint

$$\frac{1}{2}.x_1 + x_2 \leq \frac{11}{3}$$

or

$$\frac{1}{2}.x_1 + x_2 + s_1 = \frac{11}{3}$$

$$x_1, x_2, s_1 \geq 0 \text{ and integers.}$$

It may be noted that this equation can have a feasible integer solution in x_1 and x_2 only if s_1 is non-integer. This situation can be avoided in two ways:

1. The non-integer coefficients in the constraint equation can be removed by multiplying both sides of the constraint with a proper constant. For

example, the given constraint above is multiplied by 6 to obtain: $3x_1 + 6x_2 \leq 22$. However, this type of conversion is possible only when magnitudes of the integer coefficients are small.

2. Use a special cut called *Gomory's mixed-integer* cut or simply mixed-cut where only a subset of variables may assume integer values and the remaining variables (including slack and surplus variables) remain continuous. The details of developing this cut are presented below.

Method for Constructing Additional Constraint (Cut)

Consider the following mixed integer programming problem:

Maximize $Z = c_1x_1 + c_2x_2 + \ldots + c_nx_n$

subject to the constraints

$$a_{11}\,x_1 + a_{12}\,x_2 + \ldots + a_{1n}\,x_n = b_1$$
$$a_{21}\,x_1 + a_{22}\,x_2 + \ldots + a_{2n}\,x_n = b_2$$
$$\vdots \qquad\qquad\qquad \vdots$$
$$a_{m1}\,x_1 + a_{m2}\,x_2 + \ldots + a_{mn}\,x_n = b_m$$

and $\quad x_j$ are integers; $j = 1, 2, \ldots, k$ $(k < n)$.

Suppose that the basic variable x_r is restricted to be integer and has a largest fractional value among all those basic variables which are restricted to take integer values. Then rewrite the rth constraint (row) from the optimal simplex table as follows [same as Eqn. (2)]

$$x_{Br} = x_r + \sum_{j \neq r} a_{rj}x_j \qquad \ldots(1)$$

where x_j represents all the non-basic variables in the rth row except variable x_r and x_{Br} is the non-integer value of variable x_r.

Decompose coefficients of x_j, x_r variables, and x_{Bi} into integer and non-negative fractional parts as shown below:

$$x_{Br} = [x_{Br}] + f_r$$

and $\quad R_+ = \{j : a_{rj} \geq 0\}$ set of subscripts j (columns in simplex table) for which $a_{rj} > 0$

$R_- = \{j : a_{rj} < 0\}$ set of subscripts j (columns in simplex table) for which $a_{rj} < 0$

Then Eqn. (1) can be rewritten as:

$$[x_{Br}] + f_r = (1 + 0)\,x_r + 2\,a_r + \sum_{j \in R_+} a_{rj}x_j + \sum_{j \in R_-} a_{aj}x_j \qquad \ldots(2)$$

Rearrange the terms in Eqn.(2) so that all of the integer coefficients appear on the right-hand side. This gives

$$\sum_{j\in R_+} a_{rj}x_j + \sum_{j\in R_-} a_{aj}x_j = f_r + \{[x_{Br}] - x_r\} = f_r + I \qquad ...(3)$$

where f_r is a strictly positive fraction number (*i.e.,* $0 < f_r < 1$), and I is the integer value.

Since the terms in the bracket on the right-hand side of Eqn.(3) are integers, left-hand side in Eqn. (3) is either positive or negative according as $f_r + I$ is positive or negative.

Case 1 Let $f_r + I$ be positive. Then it must be $1 + f_r, + 2 + f_r, ...$ and then we shall have

$$\sum_{j\in R_+} a_{rj}x_j + \sum_{j\in R_-} a_{rj}x_j \geq f_r \qquad ...(4)$$

Since $a_{rj} \in R_-$ are non-positive and $x_j \geq 0$.

$$\sum_{j\in R_+} a_{rj}x_j + \sum_{j\in R_+} a_{rj}x_j + \sum_{j\in R_-} a_{rj}x_j$$

and hence $\sum_{j\in R_+} a_{rj}x_j \geq f_r$

Case 2 Let $f_r + I$ be negative. Then it must be $f_r, -1 + f_r, -2 + f_r, ...$ and we shall have

$$\sum_{j\in R_-} a_{rj}x_j \leq \sum_{j\in R_+} a_{rj}x_j + \sum_{j\in R_-} a_{rj}x_j \leq -1 + f_r \qquad ...(5)$$

Multiplying both sides of Eqn. (10) by the negative number $\left(\frac{f_r}{f_r - 1}\right)$, we get

$$\left(\frac{f_r}{f_r - 1}\right), \sum_{j\in R_-} a_{rj}x_j \geq f_r \qquad ...(6)$$

Either of inequalities (4) and (6) holds, since in both the cases the left-hand side is non-negative and one of these is greater than or equal to f_r. Thus, any feasible solution to mixed-integer programming must satisfy the following inequality:

$$\sum_{j\in R_+} a_{rj}x_j + \left(\frac{f_r}{f_r - 1}\right) \sum_{j\in R_-} a_{rj}x_j \geq f_r \qquad ...(7)$$

Inequality (7) is not satisfied by the optimal solution of the LP problem without integer requirement.

This is because by putting $x_j = 0$ for all j, the left-hand side becomes zero, and right-hand side becomes positive. Thus, inequality (7) defines a cut.

Adding a non-negative slack variable we can rewrite Eqn. (7) as

$$s_g = -f_r + \sum_{j\in R_+} a_{rj}x_j + \left(\frac{f_r}{f_r - 1}\right)\sum_{j\in R_-} a_{rj}x_j \qquad ...(8)$$

Equation (8) represents the required *Gomory's cut.*

For generating the cut (8) it was assumed that the basic variable x_r should take integer value. But if one or more x_j, $j \in R$ are restricted to be integers, we can proceed as follows to improve a cut shown as Eqn. (7) to a better cut, *i.e.,* the coefficients a_{rj}, $j \in R_+$ and a_{rj} $\{f_r/(f_r - 1)\}$, $j \in R_-$ are desired to be as small as possible. The value of the coefficients of x_r can be increased or decreased by an integral amount in Eqn. (8) so as to get a term with smallest coefficients in Eqn. (7). Because in order to reduce the feasible region as much as possible through cutting planes, the coefficients of integer variable x_r must be as small as possible. The smallest positive coefficient for x_r. in Eqn. (3) is:

$$\left\{f_{rj};\ \frac{f_r}{1 - f_r}(1 - f_{rj})\right\}$$

The smaller of the two coefficients would be considered to make the cut (8) penetrate deeper into the original feasible region. A cut is said to be *deep* if the intercepts of the hyperplane represented by a cut with the x-axis are larger. Obviously,

$$f_r \le \frac{f_r}{1 - f_r}(1 - f_{rj});\ f_{rj} \le f_r$$

and $$f_r > \frac{f_r}{1 - f_r}(1 - f_{rj});\ f_{rj} > f_r$$

Thus, the new cut can be expressed as

$$s_g = -f_r + \sum_{j\in R} f_{rj}^* x_j \qquad ...(9)$$

; $a_{rj} \ge 0$ and x_j non-integer

; $a_{rj} < 0$ and x_j non-integer

; $f_{rj} \le f_r$ and x_j integer

; $f_{rj} > f_r$ and x_j integer

Steps of Gomory's Mixed-Integer Programming Algorithm

Gomory's mixed-integer cutting plane method can be summarized in the following steps:

Step 1. Initialization Formulate the standard integer LP problem. Solve it by simplex method ignoring integer requirement of variables.

Step 2. Test of optimality (a) Examine the optimal solution. If all integer restricted basic variables have integer values, then terminate the procedure. The current optimal solution obtained in Step 1 is the optimal basic feasible solution to the integer LP problem.

(b) If all integer restricted basic variables are not integers, then go to Step 3.

Step 3. Generate cutting plane Choose a row r corresponding to a basic variable x_r which has largest fractional value f_r and generate a cutting plane as explained earlier in the form

$$s_g = -f_r + \sum_{j\in_+} a_{rj} x_j + \left(\frac{f_r}{f_r - 1}\right) \sum_{j\in_-} a_{rj} x_j$$

where $0 < f_r < 1$.

Step 4. Obtain the new solution Add the cutting plane generated in Step 3 to the bottom of the optimal simplex table as obtained in Step 1. Find a new optimal solution by using dual simplex method and return to Step 2. The process is repeated until all restricted basic variables are integers.

SOLVED EXAMPLES

Example 1:

Solve the following integer programming problem using Gomory's cutting plane algorithm.

Maximize $Z = x_1 + x_2$

subject to the constraints

$$3x_1 + 2x_2 \leq 5$$

$$x_2 \leq 2$$

and $x_1, x_2 \geq 0$ and are integers.

Solution:

Step 1. Obtain the optimal solution to the LP problem ignoring the integer restriction by the simplex method. The optimal solution is shown in Table 7.3.

Table 7.3 : Optimal Non-integer Solution

		$c_j \rightarrow$	1	1	0	0
c_B	Valuables in Basis B	Solution Values b (= x_B)	x_1	x_2	s_1	s_2
1	x_1	1/3	1	0	1/3	–2/3
1	x_2	2	0	1	0	1
Z = 7/2	$c_j - z_j$	0	0	–1/3	–1/3	

In Table 7.3, since all $c_j - z_j \leq 0$, the optimal solution is: x_1 = 1/3, x_2 = 2 and Max Z = 7/2.

Step 2. In the current optimal solution, all basic variables in the basis are not integers and the solution is not acceptable. Since both decision variables x_1 and x_2 are assumed to take on integer value, a pure integer cut is developed under the assumption that all the variables are integers, as explained in Step 3.

Step 3. Since x_1 is the only basic variable whose value is a non-negative fraction, we shall consider the first row for generating the Gomory cut. Considering x_1-equation as the *source row* in Table 7.3, we write

$$\frac{1}{3} = x_1 + 0.x_2 + \frac{1}{3}s_1 - \frac{1}{3}s_2 \qquad (x_1\text{-source row})$$

The factoring of the x_1-source row yields

$$\left(0 - \frac{1}{3}\right) = (1+0)x_1 + \left(0 + \frac{1}{3}\right)s_1 + \left(-1 + \frac{1}{3}\right)s_2$$

Notice that each of the non-integer coefficients is factored into integer and fractional parts in such a manner that the fractional part is strictly positive.

Rearrange the equation so that all of the integer coefficients appear on the left-hand side. This gives

$$\frac{1}{3} + (s_2 + x_1) = \frac{1}{3}s_1 + \frac{1}{3}s_2$$

Since x_1 and s_2 are non-negative, left-hand side must satisfy

$$\frac{1}{3} \leq \frac{1}{3}s_1 + \frac{1}{3}s_2 \qquad \text{(as per Eqn. 4)}$$

$$\frac{1}{3} + s_{g1} = \frac{1}{3}s_1 + \frac{1}{3}s_2$$

or $$-\frac{1}{3} = s_{g1} - \frac{1}{3}s_1 - \frac{1}{3}s_2 \qquad \text{(cut 1)}$$

where s_{g1} is the new non-negative (integer) slack variable.

By adding the Gomory cut at the bottom of Table 7.3, the new table so obtained is shown in Table 7.4.

Table 7.4

		$c_j \rightarrow$	1	1	0	0	0
c_B	Valuables in Basis B	Solution Values $b\ (= x_B)$	x_1	x_2	s_1	s_2	s_{g1}
1	x_1	1/3	1	0	1/3	–2/3	0
1	x_2	2	0	1	0	10	0
0	s_{g1}	–1/3	0	0	$\leftrightarrow$	–1/3	1→
Z = 7/2		$c_j - z_j$	0	0	–1/3	–1/3	0
	Ratio:	$(c_j - z_j)/y_{3i}$ (< 0)	–	–	1	1	–

Step 4. Apply the dual simplex method to find the new optimal solution. The key row and key column are marked in Table 7.4. The new solution is shown in Table 7.5.

Table 7.5 : Optimal Integer Solution

		$c_j \rightarrow$	1	1	0	0	0
c_B	Valuables in Basis B	Solution Values $b\ (= x_B)$	x_1	x_2	s_1	s_2	s_{g1}
1	x_1	0	1	0	0	–1	1
1	x_2	2	0	1	0	1	0
0	s_{g1}	1	0	0	1	1	–3
Z = 2		$c_j - z_j$	0	0	0	0	–1

The solution given in Table 7.5 is: $x_1 = 0$, $x_2 = 2$, $s_{g1} = 1$ and Max $Z = 2$. This also satisfies the integer requirement.

Example 2:

Maximize $\qquad Z = 12x_1 + 32x_2$

Subject to $\qquad 6x_1 + 14x_2 \leq 65$

$$5x_1 - 2x_2 \le 20$$
$$x_1, x_2 \ge 0$$
$$x_1, x_2 \text{ are integers.}$$

Solution:

Neglecing the integer constraints, the optimal solution to the problem is $x_1 = 5$, $x_2 = 2\frac{1}{2}$ with Z = 140. However, if pure integer solution is required, then x_2 may be rounded off to either 2 or 3, the resulting solution with 3 becomes infeasible. The nearest pure-integer feasible solution to (5, 2½) is (4. 2) with the corresponding Z = 112. However, the correct optimal integer solution is (0, 4) with the value of objective function Z = 128. This is a simple problem of IP and can be solved easily but in more complex problems, this is usually not possible. Thus, this indicates the requirement for special algorithms for solving Integer Programming Problem.

Example 3:

A firm manufacturing bicycles has specialised in two types of bicycles: Simple model and delux model, both requiring two types of parts (raw materials). A simple model requires 12 and 14 units of first and the second type of raw materials, while the delux model requires 28 units and 14 units of first and second kind of raw materials. In a day, the manufacturer has a supply of 84 and 70 units of the two types of raw materials. Each unit of simple model contributes a profit of Rs. 200, while the delux model contributes Rs. 320. Find the optimum number of two types of bicycles to be produced in a day to maximize the total profit.

Solution:

Above problem in the form of integer Programming Problem can be formulated as following:

If x_1 = number of simple model bicycles

x_2 = number of deluxe model bicycles

Maximize $Z = 200x_1 + 320x_2$

Subject to

$$12x_1 + 28x_2 \le 84$$
$$14x_1 + 14x_2 \le 70$$
$$x_1, x_2 \ge 0 \text{ and are integers.}$$

When computed as a Linear Programming Problem, the graphic approach gives a solution with $x_1 = 3.5$ and $x_2 = 1.5$ and the value of the objective function is Rs. 1180. However, as this solution is non-integer, it is not acceptable. If the values are rounded off to integer values, it will result in $x_1 = 3$ and $x_2 = 1$, rounding it off to next higher values will result into infeasible solution, and the value of the objective function is Rs. 920. However, this is not an optimal solution. In order to identify the integer optimum solution to the problem, moving from the non-integer optimum point (point B) towards the origin with slope of the isoprofit function, will give the optimum integer solution as the first integer point touched with the iso-profit function. This point is C, where $x_1 = 5$ and $x_2 = 0$ and the profit is 1000. It may be seen that the integer optimum solution yields a lower profit than a non-integer optimum solution. This cost is attached with the additional constraint of integer values to the decision variables.

Example 4:

Considering the same problem as considered in Example 3.

Solution:

Maximise $9x_i + 7x_2$

Subject to

$$3x_1 - x_2 \le 6$$
$$x_1 + 7x_2 \le 35$$

x_1, x_2 are non-negative integers.

The optimal continuous solution is given in Table 2.6.

Table 2.6

Basic	x_1	x_2	*Non-basic* x_3	x_4	*Solution*
x_2	0	1	−1/22	3/22	9/2
x_1	1	0	7/22	1/22	7/2
Z	0	0	28/11	15/11	63

Since the solution is non-integer, a fractional cut is added to the Table 2.6. Generally, any of the constraint equations corresponding to a non-integer solution can be selected to generate the cut. However, as a rule of thumb, we usually choose the equation corresponding to $\max_i (f_i)$. Since both equations in this problem have the same value of f_i, *i.e.*, $= f_1 = f_2 = 1/2$, either one may be taken. Considering the x_1 equation, this gives

$$x_1 + 7/22\ x_3 + 1/22\ x_4 = 7/2$$

or $$x_1 + (0 + 7/22)\ x_3 + (0 + 1/22)\ x_4 = (3 + 1/2)$$

Here the corresponding fractional cut is given by

$$S_1 - 7/22\ x_3 - 1/22\ x_4 = -1/2$$

This gives new tableau as shown in Table 2.7

Table 2.7

Basic	x_1	x_2	*Non-basic* x_3	x_4	s_1	*RHS*
x_2	0	1	–1/22	3/22	0	9/2
x_1	1	0	7/22	1/22	0	7/2
s_1	0	0	–7/22	–1/22	1	3
Z	0	0	28/11	15/11	0	63

The dual simplex method yields as in Table 2.8

Table 2.8

Basic	x_1	x_2	*Non-basic* x_3	x_4	s_1	*Solution*
x_2	0	1	0	0	–1/7	$4\frac{4}{7}$
x_1	1	0	0	1/7	1	3
x_3	0	0	1	1/7	–22/7	$1\frac{4}{7}$
Z	0	0	0	1	8	59

Since the solution is still non-integer, a new cut is generated. The x_2 equation is written as,

$$x_2 + (0 + 1/7)\ x_4 + (-1 + 6/7)S_1 + (4 + 4/7)$$

which gives the cut

$$S_2 - 1/7\ x_4 - 6/7\ S_1 = -4/7$$

Adding this constraint to the Table 2.8, the computations are obtained as in Table 2.9.

Table 2.9

Basic	x_1	x_2	x_3	Non-basic x_4	s_1	s_2	RHS
x_2	0	1	0	0	–1/7	0	$4\frac{4}{7}$
x_1	1	0	0	1/7	1	0	3
x_3	0	0	1	1/7	–22/7	0	$1\frac{4}{7}$
s_2	0	0	0	–1/7	–6/7	1	–4/7
Z	0	0	1	8	0		

The dual simplex method now yields the outcome as in Table 2.10

Table 2.10

Basic	x_1	x_2	Non-basic x_3	x_4	s_1	s_2	RHS
x_2	0	1	0	0	–1	1	4
x_1	1	0	0	0	1	0	3
x_3	0	0	1	0	–4	1	1
x_4	0	0	0	1	6	–7	4
Z	0	0	0	0	2	7	55

Which gives the optimal solution Z = 55, x_1 = 3, x_2 = 4.

Example 5:

Consider the following Integer Linear Programming

Maximize $7x_1 + 5x_2$

Subject to

$$x_1 + x_2 \leq 7$$
$$2x_1 + 6x_2 \leq 61$$
$$x_1, x_2 \geq 0 \text{ and integer.}$$

Solution

The integral linear programming solution of the problem is the integral linear programming solution is shown by dots and the associated linear programming solution space (LPO) is defined by leaving the integer constraints. The optimum LPO solution from x_1 = 4.75, x_2 = 2.25 and Z = 44.5.

The Branch and Bound procedure is based on dealing with linear programming problem only. The optimum linear programming solution of the above (x_1 = 4.75, x_2 = 2.25 and Z = 44.5) does not satisfy the integer requirements and, thus, the branch and bounding algorithm involves "modifying" the linear programming solution space in a way to allow the identification of the integer linear programming optimum. It can be observed and verified that the objective function value of any integer solution to the problem has to be less than 44.5. The observation/verification can be summarized as below:

1. In general, the integer restriction only leads to worsening of the objective function value. As such, for a maximization problem, the linear programming optimal solution is an upper bound, whereas for a minimization case, it gives the lower bound.
2. Next, the optimal solution of the linear programming as obtained above is examined. The non-integer variable is then selected (if only one is there). In case of more than one variables are non-integer, any one among them can be chosen. In the above example, x_1 and x_2 both are non-integers. For an integer solution, the value of x_1 greater than 4 and less than 5 is ruled out, and thus it can be specified that $x_1 \leq 4$ or $x_1 \geq 5$. These can be used to partition the solution space of the original linear programming into two distinct problems as follows:

P-1

Maximize $7x_1 + 5x_2$

Subject to

$$x_1 + x_2 \leq 7$$
$$10x_1 + 6x_2 \leq 61$$
$$x_1 \leq 4$$

x_1, x_2 are non-negative

P-2

Maximize $7x_1 + 5x_2$

Subject to

$$x_1 + x_2 \leq 7$$
$$10x_1 + 6x_2 \leq 61$$
$$x_1 \leq 5$$

x_1, x_2 are non-negative

The above partitioning does not rule out any integer solution to the original problem. Now, if both the problems are solved, the various possible outcomes are listed below:

(i) Both the problems have feasible solution. Let the objective function values obtained for *P*–1 and *P*–2 be *Z*–1 and *Z*–2, respectively. However, it may be noted that both these values (of *Z*–1 and *Z*–2) have to be less than or equal to *Z*, as the objective function value can never improve due to an additional constraint. The solution, thus, may indicate any of the following three outcomes:

(a) Both *P*–1 and *P*–2 have integer solution.

(b) Only one of the problems gives an integer solution.

(c) None of the problems gives integer solution.

In case (a), max. (*Z*–1, *Z*–2) can be found out, and the optimal solution is the one showing the maximum.

In case of (b), the integer solution gives a feasible solution to the integer programming problem, say *P*–1 gives an integer solution. Thus, from *P*–1, no further branching is necessary, for branching from *P*–2, first the maximum objective function value arising from *P*–2 noted, say it is *Z*–2. If *Z*–1 ≥ *Z* – 2, there is no need to explore *P*–2 further, as there cannot be any gain in doing so and the solution corresponding to *P*–1 is the required optimal solution. However, if *Z*–1 < *Z*–2, then *P*–2 has to be branched as the possibility of reaching a feasible integer solution with objective function value > *Z*–1 cannot be ruled out. Any one of the non-integer variables may be chosen, and branching can be carried out as shown earlier.

In case of (c), there is a need to branch from both the methods in exactly the same way as shown earlier. The new upper bounds for the branches from *P*–1 and *P*–2 are *Z*-1 and *Z*–2 respectively.

(ii) Both the problems are infeasible. As the whole solution space has been covered, it can be said that integer programming does not haver any feasible solution.

(iii) Only one of the problems is feasible. If the solution is integer, it would be the required one. When the solution is not integer, branching is done from the point with the upper new bound as the objective function value of the feasible problem (*P*–1 and *P*–2).

In the above example, the solution may be obtained as follows:

Step-1

Optimal solution as liner programming is determined as,

$$x_1 = 4.75,\ x_2 = 2.25, \text{ and } Z = 44.5$$

Step-2

Since the solution is not integer valued, step-2 is followed and x_1 is chosen, since $[x^*_1] = [19/4] = 4$ and the two problems are formed as,

$$LP_1 \ x_1 \leq 4$$

$$LP_2 \ x_1 \geq 5$$

LP_1 is solved and LP_2 can be represented graphically. It can observed that the two spaces contain the same integer feasible points of the integer linear programming model, *i.e.,* with respect to integer linear programming problem dealing with LP_1 and LP_2 is the same as dealing with the original linear programming space. The main difference is that the new bounding constraints ($x_1 \leq 4$ and $x_1 \geq 5$) will improve the chance of forcing the optimum extreme points of LP_1 and LP_2 toward satisfying the integer constraints. Further, the fact of bounding constraints being in the "immediate" vicinity of the continuous Linear programming space, optimum will improve their chances of giving "good" integer solution.

As the new restrictions $x_1 \leq 4$ and $x_1 \geq 5$ are mutually exclusive, LP_1 and LP_2 should be solved as two separate linear programming problems.

Step-3

LP_1 is solved and it is found that it has a feasible solution,

$$x_1 = 4, \ x_2 = 3 \text{ and } Z = 43 \qquad [A]$$

This solution satisfies the integer constraints and hence it is recorded at this step and $Z = 43$ is taken as lower bound.

Solution of LP_2 gives

$$x_1 = 5, \ x_2 = 11/6$$

and $$Z = 44\frac{1}{6}$$

Since this solution is not integer-valued, step-4 is followed.

Step-4

Now x_2 is selected. Since

$$[x^*_2] = [11/6] = 1$$

Thus the following problems are formed:

Maximize $7x_1 + 5x_2$

Subject to

$$x_1 + x_2 \leq 7$$

$$10x_1 + 6x_2 \leq 61$$

$$LP_3 \; x_1 \geq 5, x_2 \leq 1$$

$$LP_4 \; x_1 \geq 5, x_2 \geq 2$$

x_1, x_2 are non-negative

LP_3 is selected and the solution gives,

$x_1 = 5$, $x_2 = 1$ and $Z = 40$ [B]

This solution satisfies the integer conditions and thus it is recorded.

LP_4 has no feasible solution

Now, there is no more problem to be formed and thus the process ends.

At the time of ending process, it is observed that only two feasible integer solutions [A] and [B] have been noted. The "best one" of these two feasible solutions gives us the required optimum solution to the given integer programming problem.

Thus finally, the optimum solution to the given integer programming problem is,

$$x_1 = 4,\; x_2 = 3 \text{ and } Z^* = 43$$

The tree diagram (Branch and Bound Diagram) to this problem can be shown as in Fig. 2.3.

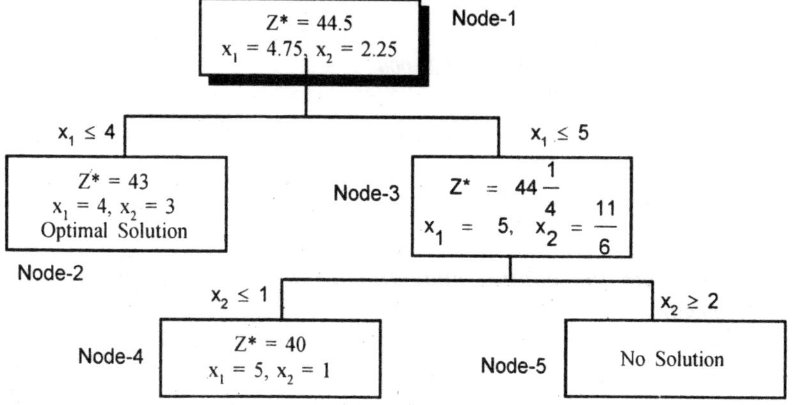

Fig 2.3 : Tree Diagram for the Branch and Bound Method.

The entire calculations of this tree diagram may be summarized in the Table 2.11.

Table 2.11

Node Problem	Solution x_1	x_2	Z^*	Additional Constraints	Type of Solution
(1)	4.75	2.25	44.5	–	Optimal Problem
(2)	4	3	43	$x_1 \leq 4$	Integer Optimal
(3)	5	11/6	$44\frac{1}{6}$	$x_1 \geq 5$	Non Integer
(4)	5	1	40	$x_1 \geq 5,\ x_2 \leq 1$	Integer
(5)	–	2	–	$x_1 \geq 5,\ x_2 \geq 2$	No Solution

Example 6:

There are four projects under consideration. Assume that the projects run into three years. Total available funds are Rs. 150,000 (to be used at the rate of Rs. 50,000 each years). The expected profit and cost break-up are as follows:

Project	*Expected Profit*		*Cost*
	Year -1	*Year-2*	*Year-3*
	1	*1,80,000*	*16,000*
20,000	*24,000*		
	2	*1,20,000*	*4,000*
10,000	*16,000*		
	3	*3,60,000*	*30,000*
20,000	*10,000*		
	4	*2,00,000*	*20,000*
10,000	*10,000*		

Formulate the capital budgeting problem as zero-one integer programming.

Solution:

Let

$$x_j = 1. \text{ if project is accepted}$$
$$= 0 \text{ otherwise, for } j = 1, 2, 3, 4$$

Then the above problem can be reduced mathematically as following:

Maximize $Z = 1{,}80{,}000x_1 + 1{,}20{,}000x_2 + 3{,}60{,}000x_3 + 2{,}00{,}000x_4$

Subject to

$$16{,}000x_1 + 4{,}000x_2 + 30{,}000x_3 + 20{,}000x_4 \le 50{,}000$$
$$20{,}000x_1 + 10{,}000x_2 + 20{,}000x_3 + 10{,}000x_4 \le 50{,}000$$
$$24{,}000x_1 + 16{,}000x_2 + 10{,}000x_3 + 10{,}000x_4 \le 50{,}000$$
$$x_1 = 0 \text{ or } 1,\ j = 1, 2, 3, 4$$

Developing an integer programming solution to the problem, the optimal solution is $x_1 = 1$, $x_2 = 1$, $x_3 = 1$ and $x_4 = 0$. The value of the objective function is Rs. 6,60,000. In other words, the projects 1, 2, and 3 are accepted for investment and the optimal expected profit would be Rs. 6,60,000.

Example 7:

A corporation is considering four possible investment opportunities. The following table presents information about the investment (in Rs thousand) profits:

Project	*Present Value of Expected Return*	*Capital Required Year-wise by Projects* Year 1	*Year 2*	*Year 3*
1	650	700	550	400
2	700	850	550	350
3	225	300	150	100
4	250	350	200	ù
Capital available for investment		1,200	700	400

In addition, projects 1 and 2 are mutually exclusive and project 4 is contingent on the prior acceptance of project 3. Formulate an integer programming model to determine which projects should be accepted and which should be rejected to maximize the present value from accepted projects.

Solution:

Let

$$x_j = \begin{cases} 1 \text{ if project j is accepted} \\ 0 \text{ if project j is rejected} \end{cases}$$

Integer LP model

Maximize (Total present value) $Z = 650x_1 + 700x_2 + 225x_3 + 250x_4$

subject to the constraints

(i) Expenditure in year 1, 2 and 3

$$700x_1 + 850x_2 + 300x_3 + 350x_4 \leq 1,200$$
$$550x_1 + 550x_2 + 150x_3 + 200x_4 \leq 700$$
$$400x_1 + 350x_2 + 100x_3 \leq 400$$
$$x_1 + x_2 \geq 1$$
$$x_4 - x_3 \leq 1$$

and $x_j = 0$ or 1.

Example 8:

Consider the following production data:

Product	*Profit per Unit (Rs.)*	*Direct Labour Requirement (hours)*
1	*8*	*15*
2	*10*	*14*
3	*7*	*17*

Fixed Cost (Rs.)	*Direct Labour Requirement*
10,000	*up to 20,000 hours*
20,000	*20,000-40,000 hours*
30,000	*40,000-70,000 hours*

Formulate an integer programming problem to determine the production schedule so as to maximize the total net profit.

Solution: Let,

x_1, x_2 and x_3 = number of units of product 1, 2 and 3, respectively to be produced

y_j = fixed cost (in Rs); j = 1,2,3.

Integer LP model

Maximize $Z = 8x_1 + 10x_2 + 7x_3 - 10,000y_1 - 20,000y_2 - 30,000y_3$

subject to the constraints

$$15x_1 + 14x_2 + 17x_3 \leq 20,000y_1 + 40,000y_2 + 70,000y_3$$
$$y_1 + y_2 + y_3 = 1$$

and $y_j = 0$ or 1

$x_j \geq 0$. 7 = 1, 2, 3.

Example 9:

Maximize $Z = 12x_1 + 32x_2$

Subject to $6x_1 + 14x_2 \leq 65$

$5x_1 - 2x_2 \leq 20$

$x_1, x_2 \geq 0$

x_1, x_2 *are integers.*

Solution:

The optimal solution to the problem is $x_1 = 5$, $x_2 = 2½$ with $Z = 140$. However, if pure integer solution is required, then x_2 may be rounded off to either 2 or 3, the resulting solution with 3 becomes infeasible. The nearest pure-integer feasible solution to (5, 2½) is (4, 2) with the corresponding Z = 112. However, the correct optimal integer solution is (0, 4) with the value of objective function Z = 128. This is a simple problem of IP and can be solved easily but in more complex problems, this is usually not possible. Thus, this indicates the requirement for special algorithms for solving Integer Programming Problem.

EXERCISES

1. Solve the following all integer programming problems, using Gomory's cutting plane algorithm:

(i) Max $Z = x_1 + 2x_2$

subject to $2x_2 \leq 2$

$x_1 + x_2 \geq 7$

$2x_1 \geq 11$

and $x_1, x_2 \geq 0$ and integers.

(ii) Max $Z = 2x_1 + 1.7x_2$

subject to $4x_1 + 3x_2 \leq 7$

$x_1 + x_2 \leq 4$

and $x_1, x_2 \geq 0$ and integers.

(iii) Max $Z = 3x_1 + 2x_2 + 5x_3$

subject to $5x_1 + 3x_2 + 7x_3 \leq 28$

$4x_1 + 5x_2 + 5x_3 \leq 30$

and $x_1, x_2, x_3 \geq 0$ and integers.

(iv) Max $Z = 3x_1 + 4x_2$

subject to $3x_1 + 2x_2 \le 8$

$x_1 + 4x_2 \ge 10$

and $x_1, x_2 \ge 0$ and integers.

(v) Max $Z = 4x_1 + 3x_2$

subject to $3x_1 + 2x_2 \le 4$

$2x_1 + x_2 \le 6$

and $x_1, x_2 \ge 0$ and integers.

(vi) Max $Z = 7x_1 + 9x_2$

subject to $-x_1 + 3x_2 \le 6$

$7x_1 + x_2 \le 35$

and $x_1, x_2 \ge 0$ and integers.

2. Solve the following integer programming problems using Gomory's Cutting Plane Algorithm or by Branch or Bound Method:

(i) Max $Z = 3x_1 + x_2$

subject to $2x_1 + 5x_2 \ge 16$

$6x_2 + 5x_2 \le 30$

and $x_2 \ge 0$

x_1 non-negative integers.

(ii) Max $Z = 4x_1 + 3x_2 + 5x_3$

subject to $2x_1 - 2x_2 + 4x_3 \ge 7$

$2x_1 + 6x_2 - 2x_3 \ge 5$

and $x_2 \ge 0$

and x_1, x_3 non-negative integers.

(iii) Max $Z = 110x_1 + 100x_2$

subject to $6x_1 + 5x_2 \le 29$

$4x_1 + 14x_2 \le 48$

and $x_1, x_2 \ge 0$ and integers.

(iv) Max $Z = 1.5x_1 + 3x_2 + 4x_3$

subject to $2x_1 + 4x_2 - x_3 \le 7$

and $x_1, x_2, x_3 \ge 0$

(v) Max $Z = 2x_1 + 3x_2$

subject to $x_1 + 3x_2 \le 9$

$$3x_1 + x_2 \leq 7$$
$$x_1 - x_2 \leq 2$$
and $x_1, x_2 \geq 0$ and integers.

(vi) Max $Z = 7x_1 + 6x_2$

subject to $2x_1 + 3x_2 \leq 12$
$$6x_1 + 5x_2 \leq 30$$
and $x_1, x_2 \geq 0$ and integers.

(vii) Max $Z = 5x_1 + 4x_2$

subject to $x_1 + x_2 \geq 2$
$$5x_1 + 3x_2 \leq 15$$
$$3x_1 + 5x_2 \leq 15$$
and $x_1, x_2 \geq 0$ and integers.

(viii) Max $Z = 3x_1 + x_2 + 3x_3$

subject to $-x_1 + 2x_2 + x_3 \leq 4$
$$2x_2 + 1.5x_3 \leq 1$$
$$x_1 - 3x_2 + 2x_3 \leq 3$$
and $x_1, x_2 \geq 0$ and integers.

3. What are the meaning and role of the lower bound and upper bound in the branch and bound method?

4. An air conditioning and refrigeration company has been awarded a contract for the air conditioning of a new computer installation. The company has to make a choice between two alternatives: (a) Hire one or more refrigeration technicians for 6 hours a day, or (b) Hire one or more part-time refrigeration apprentice technicians for 4 hours a day. The wage rate of refrigeration technicians is Rs. 400 per day, while the corresponding rate for apprentice technicians is Rs. 160 per day. The company wants to engage the technicians on work for not more than 25 man hours per day and also limit the charges to technicians to Rs. 4,800. The company estimates that the productivity of a refrigeration technician is 8 units and that of part time apprentice technician is 3 units. Formulate and solve this problem as integer LP problem to enable the company to select the optimal number of technicians and apprentices.

5. A manufacturer of toys makes two types of toys, A and B. Processing of these two toys is done on two machines X and Y. The toy A requires two hours on machine X and six hours on machine Y. Toy B requires

four hours on machine X and five hours on machine Y. There are sixteen hours of time per day available on machine X and thirty hours on machine Y. The profit obtained on both the toys is the same, *i.e.*, Rs. 5 per toy. Formulate and solve this problem as an integer LP problem to determine the daily production of each of the two toys?

6. What is integer linear programming? How does the optimal solution of an integer programming problem compare with that of the linear programming problem?
7. What is integer linear programming? Explain the merits and demerits of "rounding-off a continuous optimal solution to an LP problem to obtain an integer solution.
8. The ABC Electric Appliances Company produces two products: Refrigerators and ranges. Production takes place in two separate departments. Refrigerators are produced in department I and ranges in department II and are sold on a weekly basis. The weekly production cannot exceed 25 refrigerators in department I and 35 ranges in department II because of the limited facilities in the departments. The company regularly employs a total of 50 workers in the departments. A refrigerator requires two man-weeks of labour and a range one man-week. A refrigerator contributes a profit of Rs. 300 and a range of Rs. 200. Formulate and solve this problem as an integer LP problem to determine the units of refrigerators and ranges that the company should produce to realize maximum profit?
9. Explain how Gomory's cutting plane algorithm works.
10. The XYZ Company produces two type's of tape recorders: A-reel-to-reel model and a cassette model, on two assembly lines. The company must process each tape recorder on each assembly line and it has found that the following time is required:

Assembly Line	*Reel-to-reel*	*Cassette*
1	6 hours	2 hours
2	3 hours	2 hours

The production manager says that line 1 will be available 40 hours per week and line 2 only 30 hours per week. After these hours of operation each line must be checked for repairs. The company realizes a profit of Rs. 300 on each reel-to-reel tape recorder and Rs. 120 on each cassette recorder. Formulate and solve this problem as an integer LP problem to determine the number of recorders of each type to be produced each week in order to maximize profit.

11. Addition of a cut makes the previous non-integer optimal solution infeasible. Explain.

12. A company manufacturing metal products is planning to buy any of the following three types of lathe machines—manual, semi-automatic and fully-automatic. The manual lathe machine costs Rs. 1,500, while the semi-automatic and fully-automatic lathe machines cost Rs. 4,000 and Rs 6,000 respectively. The company's budget for buying new machines is Rs. 1,20,000. The estimated contribution towards profit from the manual, semi-automatic and fully-automatic lathe machines are: Rs 16, Rs. 20 and Rs. 22, respectively. The available floor space allows for the installation of only 36 new lathe machines. The maintenance required on fully automatic machines is low and the maintenance department can maintain 50 fully-automatic machines in a year. The maintenance of a semi-automatic machine takes 20 per cent more time than that of a fully automatic machine and a manually-operated machine takes 50 per cent more time for maintenance of a fully-automatic machine. Formulate and solve this problem as an integer LP problem to determine the optimal number of machines to be bought.

13. A stereo equipment manufacturer can produce two models A and B of 40 and 80 watts total power each. Each model passes through three manufacturing divisions 1, 2 and 3 where model A takes 4, 2.5 and 4.5 hours each and model B takes 2, 1 and 1.5 hours each. The three divisions have a maximum of 1,600, 1,200 and 1,600 hours every month, respectively. Model A gives a contribution of Rs 400 each and B ofRs 100 each. Assuming abundant product demand, formulate and solve this problem as an integer LP problem to determine the optimal product mix and the maximum contribution.

14. A manufacturing company produces two types of screwsùmetal and wooden. Each screw has to pass through slotting and threading machines. The maximum time that each machine can be run is 150 hours per month. A batch of 50 wooden screws requires 2 min on the threading machine and 3 min on the slotting machine. Metal screws of the same batch size require 8 min on the threading machine and 2 min on the slotting machine. The profit contribution for each batch of wooden and metal screws is Re 1 and Rs. 2.50, respectively. Formulate and solve this problem as an integer LP problem to determine the optimal product mix for maximum profit contribution.

15. A dietician for a hospital is considering a new breakfast menu involving oranges and cereal. This breakfast must meet the minimum

requirements for the Vitamins A and B. The number of milligrams of each of these vitamins contained in a purchasing unit for each of these foods is as follows:

Vitamin	*Milligrams per Purchasing Unit of Food*		*Minimum Requirement (mg)*
	Oranges (doz)	*Cereal (box)*	
A	1	2	20
B	3	2	50

The cost of the food ingredients is Rs. 15 per dozen for oranges and Rs. 12.50 per box for cereal. For dietary reasons, at least one unit of each Jood type must be used in the menu plan. Formulate and solve this problem as an integer programming problem.

16. The dietician at the local hospital is planning the breakfast menu for the maternity ward patients. She is primarily concerned with Vitamin E and iron requirements in planning the breakfast. According to the State Medical Association (SMA) new mothers must get at least 12 milligrams of Vitamin E and 24 milligrams of iron from breakfast. The SMA handbook reports that a scoop of scrambled egg contains 2 milligrams of Vitamin E and 8 milligrams of iron. The SMA handbook recommends that new mothers should eat at least two scoops of cottage cheese for their breakfast. The dietician considers this as one of the model constraints. The hospital accounting department estimates that a scoop of cottage cheese costs Rs. 2 and scoop of scrambled egg also costs Rs. 2. The dietician is attempting to determine the optimum breakfast menu that satisfies all the requirements and minimizes total cost. The cook insists that he can serve foods by only full scoop, thus necessitating an integer solution. Determine the optimum integer solution to the problem.

17. What is the effect of the 'integer' restriction of all the variables on the feasible space of integer programming problem?

18. A firm is considering investment in plant modernization and plant expansion. All of these proposed projects would be completed in two years, with varying requirements of money and plant engineering. The management is willing to use the following data in selecting the best set of proposals. Three resource limitations are:

First year expenditure	Rs. 4,20,000
Second year expenditure	Rs. 4,40,000
Engineering hours	15,000 hours

Project Description	*Expenditure (000s Rs.)* 1st year	2nd year	*Net Present Value*	*Engineering Hours (00) (000s Rs.)*
• Modernize shop floor	220	0	70	50
• Build new shop floor	95	270	95	80
• Equipment for new production line	0	170	60	43
• Modernize maintenance shop	60	100	80	90
• Processing sub-contract raw material	60	220	145	45
• Install new raw material processing plant	195	0	70	30
• Buy trucks and containers	80	32	45	0

The situation requires that a new or modernized shop floor be provided. The equipment for production lines applicable only to the new shop floor. The company may not want to buy or build raw material processing facilities. Formulate the given problem as an integer linear programming problem to maximize net present value of the money.

19. A building contractor has just won a contract to build a municipal library building. His present labour workforce is inadequate to take up this work immediately as he has already got other jobs on hand and, therefore, has to decide immediately whether to hire one or more labourers on a full-time basis (eight hours a day each) or to allow over-time to one or more of the existing labour force (five hours a day each). Extra labourers can be hired for Rs. 40 per day (for eight hours) while over-time costs Rs. 43 per day (for five hours per day). The contractor wants to limit his extra payment to Rs 400 per day and to use no more than twenty labourers (both full-time and over-time) because of limited supervision. He estimates that the new labour employed on a full-time basis will generate Rs. 15 a day in profits, while this time labour Rs. 20 a day. Formulate and solve this problem as an integer LP problem to help the building contractor to decide the optimum labour force.

20. A company produces two products A and B. Each unit of product A requires one hour of engineering services and five hours of machine time. To produce one unit of product B, two hours of engineering and 8 hours of machine time are needed. There are 100 hours of

engineering and 400 hours of machine time available. The cost of production is a non-linear function of the quantity produced as given in the following table:

Product A		*Product B*	
Production (units)	*Unit Cost (Rs.)*	*Production (units)*	*Unit Cost (Rs.)*
0-50	10	0-40	7
50-100	8	40-100	3

The unit selling price of product A is Rs. 12 and of product B is Rs. 14. The company would like a production plan which gives the number of units of A and the number of units ofB to be produced that will maximize profit. Formulate and solve this problem as an integer linear programming problem to help the company to maximize the total revenue.

21. The XYZ Corporation manufactures an electric device, final assembly of which is accomplished by a small group of trained workers operating simultaneously on the devices. Due to space limitations, the working group may not exceed ten in number. The firm's operating budget allows Rs. 5,400 per month as salary for the group. A certain amount of discrimination is evidenced by the fact that the firm pays men in the group Rs. 700 per month, while women doing the same work receive Rs 400. However, previous experience has indicated that a man will produce about Rs. 1,000 in value added per month, while a woman worker adds Rs. 900. If the firm wishes to maximize the value added by the group, how many men and women should be included? (A non-integer solution for this problem will not be accepted).

22. A manufacturer of baby dolls makes two types of dolls. One is sold under the brand name 'Molina' and the other 'Suzie'. Processing of these two dolls is done on two machines A and B. The processing time for each 'Molina' is 3 hours and 4 hours on machines A and B, respectively and that for each 'Suzie' is 1 hour and 4 hours on machines A and B, respectively. There is 50 hours of time per week available on machine A and 90 hours on machine B. The profit contribution for a 'Molina' is Rs. 6 and that for a 'Suzie' is Rs. 18. Formulate and solve this problem as an integer LP problem to determine the optimal weekly production schedule of the two dolls.

23. A firm makes two products: X and Y, and has total production, capacity of 9 tonnes per day, X and Y requiring the same production

capacity. The firm has a permanent contract to supply at least 2 tonnes of X and at least 3 tonnes of Y per day to another company. Each tonne of X requires 20 machine-hours production time and each tonne of Y requires 50 machine hours production time. The daily maximum possible number of machine- hours is 350. All the firm's output can be sold and the profit made is Rs. 80 per tonne of X and Rs. 120 per tonne of Y. It is required to determine the production schedule for maximum profit and to calculate this profit. (A non-integer solution for this problem will not be accepted).

24. Describe any one method of solving a mixed-integer programming problem.
25. Discuss the advantages and disadvantages of solving integer programming problems by :
    (a) cutting plane method, and
    (b) branch and bound method.
26. The ABC company requires an output of at least 200 units of a product per day and to accomplish this target it can buy machines A or B or both. Machine A costs Rs 20,000 and B Rs. 15,000 and the company has a budget of Rs. 2,00,000 for the same. Machines A and B will produce 24 and 20 units respectively of this product per day. However, machine A will require a floor space of 12 square feet while machine B will require 18 square feet and the company has a total floor space of 180 square feet only. Formulate and solve this problem as an integer LP problem to determine the minimum number of machines that should be purchased.
27. Sketch the branch and bound method in integer programming.
28. ABC Manufacturing company faces the following problem: Should they make or buy each of several products? The company policies specify that they will either make or buy the whole lot of each product in a complete lot. The company has four products to-make or buy with six machines involved in making these products if they are made in the shop. The times per unit (in hours) required are as follows:

			Machine			
Product	***A***	***B***	***C***	***D***	***E***	***F***
1.	0.04	0.02	0.02	0	0.03	0.06
2.	0	0.01	0.05	0.15	0.09	0.06
3.	0.02	0.06	0	0.06	0.02	0.02
4.	0.06	0.04	0.15	0	0	0.05

Forty hours are available on each machine. One hundred ten units of each product are needed. The costs of making the products are listed below:

Products	:	1	2	3	4
Cost/unit (in Rs.)	:	2.25	2.22	4.50	1.90

The cost to buy the products are

Product	:	1	2	3	4
Cost/unit (in Rs.)	:	31.0	2.60	4.75	2.25

Formulate this problem as a zero-one integer programming problem.

29. Discuss the advantages of the branch and bound method.
30. An Airline corporation is considering the purchase of three types of jet planes. The purchase price would be Rs. 45 crore for each A type plane; Rs. 40 crore for each B type plane and Rs. 25 crore for each C type plane. The corporation has resources worth Rs. 500 crore for these purchases. The three types of planes, if purchased, would be utilized essentially at maximum capacity. It is estimated that the net annual profit would be Rs. 3 million for A type, Rs. 2.25 million for B type and Rs. 1.5 million for C type plane. Each plane requires one pilot and it is estimated that 25 trained pilots will be available. If only C type planes were purchased, the maintenance facilities would be able to handle 30 new planes. However, each B type is equivalent to one C type plane and each A type plane is equivalent to one C type plane, in terms of their use of the maintenance facilities.

    The management of the corporation wants to know how many planes of each type (considering the fact that number of planes should be an integer) should be purchased in order to maximize profits.
31. The owner of a ready-made garments store makes two types of shirts: Arrow and Wings. He makes a profit of Rs. 10 and Rs. 15 per shirt on Arrow and Wings, respectively. He has two tailors A and B at his disposal to stitch these shirts. Tailors A and B can devote at the most 12 hours each day. Both these shirts are to be stitched by both the tailors. Tailor A and Tailor B spend 3 hours and 4 hours, respectively in stitching an Arrow shirt and 4 hours and 3 hours, respectively in stitching a Wings shirt. How many shirts of both the types should be stitched in order to maximize daily profits? (A non-integer solution for this problem will not be accepted.)
32. The dietician at the local hospital is planning the breakfast menu for the maternity ward patients. She is planning a special non-fattening diet, and has chosen cottage cheese and scrambled eggs for breakfast. She is primarily concerned with Vitamin E and Iron requirements in planning the breakfast.

According to the State Medical Association (SMA) new mothers must get at least 12 milligrams of Vitamin E and 24 milligrams of iron from breakfast. The SMA handbook reports that a scoop of cottage cheese contains 3 milligrams of Vitamin E and 3 milligrams of iron. An average scoop of scrambled egg contains 2 milligrams of Vitamin E and 8 milligrams of iron. The SMA handbook recommends that new mothers should eat at least two scoops of cottage cheese for their breakfast. The dietician considers this as one of the model constraints.

The hospital accounting department estimates that a scoop of cottage cheese costs Re 1, and a scoop of scrambled egg also costs Re 1. The dietician is attempting to determine the optimum breakfast menu that satisfies all the requirements and minimize total cost. The cook insists that he can serve foods by only full scoop, thus necessitating an integer solution. Formulate and solve this problem as an integer LP problem to determine the optimum-integer solution to the problem.

33. Suppose five items are to be loaded on a vessel. The weight (W), volume (V) and price (P) are tabulated below. The maximum cargo weight and cargo volume are W = 112 and V = 109, respectively. Determine the most valuable cargo load in discrete unit of each item.

Item	*W*	*V*	*Price (Rs)*
1	5	1	4
2	8	8	7
3	3	6	6
4	2	5	5
5	7	4	4

Formulate and solve this problem as an integer linear programming model.

34. ABC Construction Company is faced with the problem of determining which projects it should undertake on the next 4 years. The following table gives information about each project:

Project	*Estimated Present Value*	*Captial Requirements*			
		Year 1	*Year 2*	*Year 3*	*Year 4*
A	18,00,000	3,00,000	4,00,000	4,00,000	3,00,000
B	2,00,000	1,20,000	80,000	0	40,000
C	7,20,000	3,00,000	2,00,000	2,00,000	2,00,000
D	8,00,000	2,00,000	4,00,000	4,00,000	1,00,000
	Fund available	6,50,000	8,00,000	8,00,000	5,00,000

Formulate a zero-one programming model to maximize estimated value.

35. The cutting division of Photo Films Corporation requisitions from the stock control department, plastic films of 85 feet of fixed unit length which can be cut according to two different patterns. The first pattern will cut each film length into two 35-feet pieces with the remaining 15 feet to scrap. Second pattern will cut each film length into a 35-feet piece and two 25-feet pieces with nothing-to scrap. Present order from a customer is for 8 pieces of 35 feet length and six pieces of 25 feet length. What number of plastic films of 85 feet be cut according to the patterns (assuming both patterns have to be used) in order to minimize the scrap? (A non-integer solution for this problem will not be accepted).

36. A manufacturing company produces two products, each of which requires stamping, assembly and painting operations. Total productive capacity by operation if it were devoted soley to one product or the other is:

Operation	*Productive Capacity (units per week)*	
	Product A	*Product B*
Stamping	50	75
Assembly	40	80
Painting	90	45

Pro-rata allocation of productive capacity is permissible so also combinations of production of the two products. Demand for the two products is unlimited and the profit on A and B are Rs. 150 and Rs. 120, respectively. Determine the optimal product mix. (Non-integer solution for this problem will not be accepted.) 24. Write constraints to satisfy each of the following conditions in a project selection model. The projects are numbered 1, 2, 3, 4, 5, 6, 7, 8, 9, and 10.

(i) Exactly one project from the set (1, 2, 3) must be selected.

(ii) Project 2 can be selected only if number 10 is selected. However, 10 can be selected without 2 being selected.

(iii) No more than one project from the set (1, 3, 5, 7, 9) can be selected.

(iv) If number 4 is selected, then number 8 cannot be selected.

(v) Project 4 and 10 must both be selected or both be rejected.

37. A production manager faces the problem of job allocation between his two production crews. The production rate of crew one is five units per hour and it is six units for crew two. The normal working hours for each crew is 8 hours per day. If the firm requires over-time operations each crew can work up to 11 hours (union contract). The firm has a special contract with a customer to provide a minimum of 120 units of the product the next day.

    The union contract calls for working arrangements where each crew should work at least 8 hours per day, and any over-time not exceeding 3 hours should be in terms of increments of an hour. The operating costs per hour are Rs. 200 for crew one and Rs. 220 for crew two. Formulate and solve this problem as an integer programming model to determine optimum job allocation.

38. An airline owns an ageing fleet of Boeing 737 jet airplanes. It is considering a major purchase of up to 17 new Boeing models 757 and 767 jets. The decision must take into account several cost and capacity factors including the following:

    (i) the airline can finance up to Rs. 4,000 million in purchases;

    (ii) each Boeing 757 will cost Rs. 350 million, while each Boeing 767 will cost Rs. 220 million;

    (iii) at least one-third of the planes purchased should be the longer-ranged 757;

    (iv) the annual maintenance budget is to be no more than Rs. 80 million;

    (v) the annual maintenance cost per 757 is estimated to be Rs. 8,00,000 and it is Rs. 5,00,000 for each 767 purchased;

    (vi) each 757 can carry 1,25,000 passengers per year, while each 767 can fly 81,000 passengers annually. Formulate this problem as an integer programming problem to maximize the annual passenger carrying capacity and solve it by the cutting plane method.

39. Write constraints to satisfy each of the following conditions in a project selection model. The projects are numbered 1, 2, 3, ..., 10.

    (i) Exactly one project from the set (1, 2, 3) must be selected.

    (ii) Project 2 can be selected only if number 10 is selected. However, 10 can be selected without 2 being selected.

    (iii) No more than one project from the set (1, 3, 5, 7, 9) can be selected.

(iv) If number 4 is selected, then number 8 cannot be selected.

(v) Projects 4 and 10 must both be selected or both be rejected.

40. A stereo equipment manufacturer can produce two models A and B of 40 and 80 watts total power each. Each model passes through three manufacturing divisions 1, 2 and 3 where model A takes 4, 2.5 and 4.5 hours each and model B takes 2, 1 and 1.5 hours each. The three divisions have a maximum of 1,600, 1,200 and 1,600 hours every month, respectively. Model A gives a contribution of Rs. 400 each and B of Rs. 100 each. Assuming abundant product demand, formulate and solve this problem as an integer LP problem to determine the optimal product mix and the maximum contribution.

41. A company makes two products, each of which requires time on four different machines. Only integer amounts of each can be made. Company wants to find the output mix that maximizes total profits, given the specifications shown in the following table:

Total Machine Hours Available	*Machine*	*Machine Hours (per unit)* Product I	Product II
3400	A	200	500
1450	B	100	200
3000	C	400	300
2400	D	400	100
Profit/unit (Rs. 1,000)		6	6

MULTIPLE CHOICE QUESTIONS

1. The 0-1 integer programming problem

(a) requires the decision variables to have values between zero and one

(b) requires that the constraints all have coefficients between zero and one

(c) requires that the decision variables have coefficients between zero and one

(d) all of the above

2. The part of the feasible solution space eliminated by plotting a cut contains

(a) only non-integer solutions

(b) only integer solutions

(c) both (a) and (b)

(d) none of the above

3. In the Branch and Bound approach to a Max IL problem, a node is terminated if

(a) a node has an infeasible solution

(b) a node yields a solution that is feasible but not an integer

(c) upper bound is less than the current sub-problem's lower bound

(d) all of the above

4. While solving IP problem any non-integer variable in the solution is picked up to

(a) obtain the cut constraint (b) enter the solution

(c) leave the solution (d) none of the above

5. Branch and Bound method divides the feasible solution space into smaller parts by

(a) branching (b) bounding

(c) enumerating (d) all of the above

6. Rounding off solution values of decision variables in an LP problem may not be acceptable because

(a) it does not satisfy constraints

(b) it violates non-negativity conditions

(c) objective function value is less than the objective function value of LP

(d) none of the above

7. Which of the following is the consequence of adding a new cut constraint to an optimal simplex table

(a) addition of a new variable to the table

(b) makes the previous optimal solution infeasible

(c) eliminates non-integer solution from the solution space

(d) all of the above

8. In a Branch and Bound minimization tree, the lower bounds on objective function value

(a) do not decrease in value

(b) do not increase in value

(c) remain constant

(d) none of the above.

9. In a mixed-integer programming problem

   (a) all of the decision variables require integer solutions

   (b) few of the decision variables require integer solutions

   (c) different objective functions are mixed together

   (d) none of the above

10. The use of cutting plane method

    (a) reduces the number of constraints in the given problem

    (b) yields better value of objective function

    (c) require use of standard LP approach between each cutting plane application

    (d) all of the above.

3

Dual Problem

INTRODUCTION

In the context of linear programming, duality implies that each linear programming problem can be analysed in two different ways but having equivalent solutions. Each LP problem (both maximization and minimization) stated in its original form has associated with another linear programming problem (called *dual linear programming problem* or in *short dual*), which is unique, based on the same data. In general it is immaterial which of the two problems is called *primal* or *dual*, since the dual of the dual is primal.

The problem of production planning, by using primal LP problem, the production manager attempts to optimize resource allocation by determining quantities for each product to be produced that will maximize profit. But through dual LP problem approach, he attempts to achieve a production plan that optimizes resource allocation so that each product is produced at that quantity such that its marginal opportunity cost equals its marginal return. Thus, the main focus of dual is to find for each resource its best marginal value or shadow price. This value reflects the scarcity of the resources, *i.e.,* the maximum additional prices to be paid to obtain one additional unit of the resources to maximize profit under the resource constraints. If a resource is not completely used, *i.e.,* there is slack, then its marginal value is zero.

The shadow price is also defined as the rate of change in the optimal objective function value with respect to the unit change in the availability of a resource. Precisely for any constraint, we have

$$\text{Shadow price} = \frac{\text{Change in optimal objective function value}}{\text{Unit change in the availability of resource}}$$

The format of the simplex method is such that solving one type of problem is equivalent to solving the other simultaneously. Thus, if the optimal solution to one is known, the optimal solution of the other can also be read from the $c_j - z_j$ row of the Final Simplex Table. In some cases, considerable computing time can be reduced by solving the dual.

FORMULATION OF DUAL LINEAR PROGRAMMING PROBLEM

There are two important forms of primal and dual problems, namely the symmetrical (canonical) form and the standard form.

To understand the concept of Dual problem more clearly, consider the following diet problem:

The amount of two vitamins A and B per unit present in two different foods A_1 and A_2 respectively are given below:

Vitamin	*Food*		*Minimum daily requirement*
	A_1	A_2	
A	6	9	60
B	4	13	108
Cost (per unit)	Rs. 12	Rs. 18	

The diet problem is to ascertain the quantities of foods A_1 and A_2 that should be eaten to meet the minimum daily requirements of vitamins A and B at a minimum cost.

Let x_1 and x_2 be the number of units of foods A_1 and A_2 to be purchased respectively. Then the problem is to find the values of x_1 and x_2 which minimize

$$Z_x = 12x_1 + 18x_2$$

subject to the constraints

$$6x_1 + 9x_2 \geq 60$$

$$4x_1 + 13x_2 \geq 108,$$

$$x_1, x_2, \geq 0.$$

We shall consider this linear programming problem as the primal problem. Now we shall formulate the dual corresponding to this primal.

Suppose a wholesale dealer sells two vitamins A and B. Customers purchase the two vitamins from him in the form of two foods A_1 and A_2 (as given in the above). The foods A_1 and A_2 have their market value only because of their vitamin contents. Now, the dealer has to fix-up the maximum per unit selling prices for the two vitamins A and B in such a way that the resulting prices of the foods A_1 and A_2 do not exceed their existing market prices.

Suppose the dealer decides to fix-up the two prices y_1 and y_2 respectively. Then the problem is to determine the values of y_1 and Y_2 which maximize

$$Zy = 60y_1 + 108y_2$$

subject to

$$6y_1 + 4y_2 \leq 12$$

$$9y_1 + 12y_2 \leq 18,$$

$$y_1, y_2 \geq 0.$$

Observing the above primal and the dual a little closely, we find that

(i) Primal is a minimization problem while the dual is a maximization problem.

(ii) The constraint values 60, 108 of the primal have become the coefficients of the dual variables y_1 and y_2 in the objective function of the dual in that order. The coefficients of the variables x_1 and x_2 in the objective function of the primal have become the constraint values in the dual.

(iii) The constraint coefficient matrix of the dual is the transpose of the constraint coefficient matrix of the primal.

(iv) The direction of inequalities in the dual is the reverse of that in the primal.

We can represent the primal-dual relationship in the following form:

Primal	***Dual***
Minimize	Maximize
$\overset{c}{} \quad \overset{x}{}$ $Z_x = [12 \;\; 18] \begin{bmatrix} x_1 \\ x_2 \end{bmatrix}$	$\overset{b'}{} \quad \overset{y}{}$ $Z_y = [60 \;\; 108] \begin{bmatrix} y_1 \\ y_2 \end{bmatrix}$
subject to	subject to
A x b $\begin{bmatrix} 6 & 9 \\ 4 & 13 \end{bmatrix} \begin{bmatrix} x_1 \\ x_2 \end{bmatrix} \geq \begin{bmatrix} 60 \\ 108 \end{bmatrix}$	A' y c' $\begin{bmatrix} 6 & 4 \\ 9 & 13 \end{bmatrix} \begin{bmatrix} y_1 \\ y_2 \end{bmatrix} \leq \begin{bmatrix} 12 \\ 18 \end{bmatrix}$
$x_1, x_2 \geq 0.$	$y_1, y_2 \geq 0.$

Symmetric Dual Problems

The symmetric relation between primal and dual problem exists.

Primal Problem: Consider a L.P. problem.

Mini $Z_p = c_1 x_1 + c_2 x_2 + ... + c_n x_n$

s.t. $a_{11} x_1 + a_{12} x_2 + ... + a_{1n} x_n \geq b_1$

$a_{21} x_1 + a_{22} x_2 + ... + a_{2n} x_n \geq b_2$

$a_{m1} x_1 + a_{m2} x_2 + ... + a_{mn} x_n \geq b_m$

and $x_1, x_2, x_3, ... ,x_n \geq 0.$

The dual problem of above L.P. problem is obtained by :

(i) Changing the direction of inequalities.

(ii) Maximizing instead of minimizing.

(iii) Transporting the coefficient matrix.

(iv) Interchanging the role of constant terms and the coefficients of the objective function.

The dual problem is as follows.

Max. $Z_D = b_1 w_1 + b_2 w_2 + ... + b_m w_m$

s.t. $a_{11} w_1 + a_{21} w_2 + ... + a_{m1} w_m \leq c_1$

$a_{21} w_1 + a_{22} + ... + a_{m2} w_m \leq c_2$

$a_{1n} w_1 + a_{2n} w_2 + ... + a_{mn} w_m \leq c_n$

and $w_1, w_2, ... , w_m \geq 0.$

In matrix form the primal and dual problems can be written as

Primal Problem: Find the column vector x which

Mini. $Z_p = c x$

s.t. $Ax \geq b$

and $x \geq 0.$

Dual Problem : Find a column vector w which

Maxi $Z_D = b' w$

s.t. $A' w \leq c',$

and $w \geq 0.$

The above primal and dual problem is called symmetric dual problems.

Unsymmetrical Dual Problem

Primal Problem. Find a column vector x which

Mini. $Z_p = c x$

s.t. $\quad A x = b, x \geq 0$

Dual problem. Find a column vector w which

Max. $\quad Z_D = b' w$

s.t. $\quad A' w \leq c'$.

The Dual of Mixed System : If a system consist of a monitor of equations, inequalities (in either direction), non-negative variable, or unrestricted.

RULES FOR CONSTRUCTING THE DUAL FROM PRIMAL

The rules for constructing the dual from the primal or primal from the dual when using the symmetrical form are:

1. The right hand side constants $b_1, b_2, \ldots, b_m$ of the primal becomes the coefficients of the dual variables $y_1, y_2, \ldots, y_m$ in the dual objective function Z . Also the coefficients $c_1\ c_2, \ldots, c_n$ of the primal variables $x_1\ x_2, \ldots, x_n$ in the objective function become the right hand side constants in the dual.
2. If the objective function of the primal is to be maximized, the objective function of the dual becomes minimization and *vice-versa.*
3. For a maximization primal with all ≤ (less than or equal to) type constraints, there exists a minimization dual problem with all ≤ (greater than or equal to) type constraints and *vice-versa.* Thus, the inequality sign is reversed in all the constraints except the non-negativity conditions.
4. The matrix of the coefficients of .variables in the constraints of dual is the transpose of the matrix of coefficients of variables in the constraints of primal and *vice-versa.* That is, coefficients of the primal variables $x_1, x_2, \ldots, x_n$ in the constraints of primal LP problem are the coefficients of dual variables in first, second, ..., nth, constraints for the dual problem, respectively.
5. If the rth dual variable is unrestricted in sign, then jth primal constraint is = (equality) type and *vice-versa.*
6. Each constraint in the primal corresponds to a dual variable in the dual and *vice-versa.* Thus, given a primal problem with m constraints and n variables, there exists a dual problem with m variables and n constraints.

STANDARD RESULTS ON DUALITY

1. The dual of the dual LP problem is again the primal problem.

2. If either the primal or the dual problem has an unbounded objective function value, the other problem has no feasible solution.
3. If either the primal or dual problem has a finite optimal solution, the other one also possesses the same, and the optimal value of the objective functions of the two problems are equal, *i.e.*, Max Z_x = Min Z_y. This analytical result is known as the *fundamental primal-dual* relationship.
4. Complementary slackness property of primal-dual relationship states that for a positive basic variable in the primal, the corresponding dual variable will be equal to zero. Alternatively, for a non-basic variable in the primal (which is zero), the corresponding dual variable will be basic and positive.

PRINCIPLE OF COMPLEMENTARY SLACKNESS

The principle of complementary slackness establishes the relationship between the optimal value of a primal variable in one problem with a slack or surplus variable of the dual problem. To illustrate this concept, let us consider the example of production planning.

1. If at the optimal solution of a primal problem, a primal constraint has positive value of slack variable, the corresponding resource is not completely used up and must have zero opportunity cost (shadow price). This means that having more of this resource will not improve the value of the objective function. But if the value of slack variable is zero in that constraint, the entire resource is being used and must have a positive opportunity cost.

   Slack variables in a primal LP problem correspond to the dual variables in the dual LP problem. Thus, complementary slackness states that for every resource, the following condition must hold

   $$\text{Primal slack variable} \times \text{Dual variable} = 0.$$

2. When resources are not used completely for producing any unit of a product, in that case opportunity cost of such resources exceeds the unit profit from that product. But if some of the product is made, the opportunity cost of its resources must equal its unit profit, so that surplus product cost is zero. A primal variable in primal LP problem represents quantity of product made and surplus variables in dual represent surplus product costs. In this case complementary slackness states that for every unit of the product, following condition must hold.

   $$\text{Primal variable} \times \text{Dual surplus variable} = 0.$$

MANAGERIAL SIGNIFICANCE OF DUALITY

The importance of the dual LP problem is in terms of the information which it provides about the value of the resources. The economic analysis is concerned in deciding whether or not to secure more resources and how much to pay for these additional resources.

The significance of the study of dual is as follows:

(i) The dual variables provide the decision-maker a basis for deciding how much to pay for additional units of resources.

(ii) The maximum amount that should be paid for one additional unit of a resource is called its shadow price (also called *simplex multiplier*).

(iii) The total marginal value of the resources equals the optimal objective function value. The dual variables equal the marginal value of resources (shadow prices).

The value of the ith dual variable is the rate at which the primal objective function value will increase as the ith resource increases, assuming that all other data is unchanged.

MATRIX FORM OF SYMMETRIC PRIMAL-DUAL PROBLEM

Primal Problem: Find a column vector $\mathbf{x} \in R^n$ which maximizes

$$Z_x = c\,x,\ c \in R^n$$

subject to $A\,x \leq b,\ b \in R^m$,

$$x \geq 0$$

and A is an $m \times n$ real matrix.

Dual Problem: Find a column vector $w \in R^m$ which minimizes

$$Z_w = b'\,w$$

subject to $A'\,w \geq c'$

$w \geq 0$, A', b', c'

are the transposes of A, b and c respectively.

SOME DUAL PROBLEM THEOREMS

We observe that dual of a primal problem is also a linear programming problem therefore we can also construct the dual of a dual problem.

Here we shall prove a number of fundamental theorems to describe the relation between the primal and its dual. The relationship between the primal and dual is extremely useful in the development of mathematical programming.

Theorem 1:

If any of the constraints in the primal is a strict equality, the corresponding dual variable is unrestricted in sign.

Proof:

Let L.P.P. is

Mini $Z_P = c\,x$

s.t. $\sum_{j=1}^{n} a_{ij}x_j \geq b_i \ \forall_i$

and $\sum_{j=1}^{n} a_{kj}x_j = b_k \ \forall_i \geq k$

$x_j \geq 0 \ \forall_j$

In the standard primal form the above L.P.P. can be written as

Min $Z_P = c\,x$

s.t. $\sum_{j=1}^{n} a_{ij}x_j \geq b_i \ \forall_i \geq k$

and $\sum_{j=1}^{n} a_{kj}x_j \geq b_k$

$-\sum_{j=1}^{n} a_{kj}x_j \geq -b_k$

Now the dual of the above problem becomes

Max. $Z_D = \sum_{i=1}^{m} b_i w_i + b_k (w_k' - w_k'')$

$w_i \geq 0,\ i \neq k,\ w_k',\ w_k'' \geq 0$

Now writing $w_k'' - w_k'' = w_k$, the dual problem becomes

Max. $Z_D = \sum_{i=1}^{m} b_i w_i$

s.t. $\sum_{i=1}^{m} a_{ji}w_j \leq c_j \ \forall_j = 1, 2, \ldots n$

$w_j \geq 0 \ \forall_j \geq k$

obviously w_k is unrestricted in sign since $w_k >$ or < 0 according as $w_k' >$ or w_k''.

Theorem 2:

If any variable of the primal is unrestricted in sign, the corresponding constraint in the dual will be a stick equality.

Proof:

Let the kth variable x_k of the primal be unrestricted in sign.

Therefore replacing x_k by $x_k' - x_k''$ where x_k'' and x_k'' are non-negative variables, the primal can be written as

Max. $Z_D = c_1 x_1 + c_2 x_2 + \ldots + c_k (x_k' - x_k'') + c_{k+1} x_{k+1} + \ldots + c_n x_n$

s.t. $a_{11} x_1 + a_{12} x_2 + \ldots + a_{1k} (x_k' - x_k'') + a_{k+1} x_{k+1} + \ldots + a_{1n} x_n \le b_1$

....

....

....

$a_{m1} x_1 + a_{m2} x_2 + \ldots + a_{mk} (x_k' - x_k'') + a_m, b_{11} x_{b+1} + a_{mn} x_n \le b_m$

and $x_1, \ldots, x_{k-1}, x_k', x_b'', x_{k+1}, \ldots, x_n \ge 0$

The dual of the above problem is given by

Min. $Z_D = b_1 w_1 + b_2 w_2 + \ldots + b_m w_m$

and $a_{11} w_1 + a_{21} w_2 + \ldots + a_{m1} w_m \ge c_1$

....

....

$a_{12} w_1 + a_{22} w_2 + \ldots + a_{m2} w_m \ge c_2$

$a_{1k} w_1 + a_{2k} w_2 + \ldots + a_{mk} w_m \ge c_k$

....

....

$a_{1k} w_1 - a_{2k} w_2 - \ldots - a_{mk} w_m \ge - c_k$

$a_{1n} w_1 + a_{2n} w + \ldots + a_{mn} w_m \ge c_n$

and $w_1, w_2, \ldots, w_m \ge 0$

The two constraints under bracket {} in the above dual are equivalent to an equality

$a_{1k} w_1 + a_{2k} w_2 + \ldots + a_{mk} w_m = c_k$

i.e., the kth constraint in the dual is an equality.

Difference between Primal and Dual

Primal	Dual
1. Objective function Mini ZP.	Objective function max Z_D.
2. Requirement vector	Price vector
3. Coefficient matrix	Transpose of the coefficient matrix
4. Relation	Variable
5. i^{th} inequality	i^{th} variable $w_i \geq 0$
6. i^{th} constraint an equality.	i^{th} variable w_i unrestricted in sign.
7. constants with sign $\geq$	Constants with sign $\leq$
8. Variable	
9. i^{th} variable $x_i > 0$	i^{th} relation a strict equation.
10. i^{th} variable x_i unrestricted in sign	i^{th} constraint a strict equality
11. i^{th} slack variable positive	i^{th} variable zero
12. i^{th} variable zero	i^{th} surplus variable positive
13. Finite optimal solution	Finite optimal solution with equal optimal value of the objective function.
14. Unbounded solution	No solution or an unbounded solution.

Theorem 3:

The dual of a dual of the given primal is the primal itself.

Proof:

Suppose the given linear programming problem is

Primal Problem

$$\text{Max. } Z_P = c_1 x_1 + c_2 x_2 + \ldots + c_n x_n$$

subject to

$$\begin{aligned} a_{11} x_1 + a_{12} x_2 + \ldots + a_{1n} x_n &\leq b_1 \\ a_{21} x_1 + a_{22} x_2 + \ldots + a_{2n} x_n &\leq b_2 \\ \ldots \quad \ldots \quad &\ldots \\ \ldots \quad \ldots \quad &\ldots \\ a_{m1} x_1 + a_{m2} x_2 + \ldots + a_{mn} x_n &\leq b_m, \end{aligned} \qquad \ldots(1)$$

$$x_1, x_2, \ldots, m_n \geq 0.$$

Dual Problem

The dual of the above primal can be written as

Mini. $Z_D = b_1 w_1 + b_2 w_2 + \ldots + b_m w_m \geq c_1$

subject to

$$\begin{array}{l} a_{11} w_1 + a_{21} w_2 + \ldots + a_{m1} w_m \geq c_1 \\ a_{12} w_1 + a_{22} w_2 + \ldots + a_{m2} w_m \;^{3}\; c_2 \\ \ldots \quad \ldots \quad \ldots \\ \ldots \quad \ldots \quad \ldots \\ a_{1n} w_1 + a_{2n} w_2 + \ldots + a_{mn} w_m \;^{3}\; c_n \\ w_1, w_2, \ldots, w_m \geq 0. \end{array} \quad \ldots(2)$$

Now we have to construct the dual of the above dual. First we shall change the above dual to standard maximization form.

Max. $(-Z_D) = -b_1 w_1 - b_2 w_2 - \ldots - b_m w_m$

subject to

$$\begin{array}{l} -a_{11} w_1 - a_{21} w_2 - \ldots - a_{m1} w_m \leq -c_1 \\ -a_{12} w_1 - a_{22} w_2 - \ldots - a_{m2} w_m \leq -c_2 \\ \ldots \quad \ldots \quad \ldots \\ \ldots \quad \ldots \quad \ldots \\ -a_{1n} w_1 - a_{2n} w_2 - \ldots - a_{mn} w_m \leq -c_n \\ w_1, w_2, \ldots, w_m \geq 0. \end{array} \quad \ldots(3)$$

Dual of the Dual

Considering the above dual as primal, its dual can be written as

Mini. $Z_y = -c_1 y_1 - c_2 y_2 - \ldots - c_n y_n$

subject to

$$\begin{array}{l} -a_{11} y_1 - a_{12} y_2 - \ldots - a_{1n} y_n \geq -b_1 \\ -a_{21} y_1 - a_{22} y_2 - \ldots - a_{2n} y_n \geq -b_2 \\ \ldots \quad \ldots \quad \ldots \\ \ldots \quad \ldots \quad \ldots \\ -a_{m1} y_1 - a_{m2} y_2 - \ldots - a_{mn} y_n \geq -b_m, \\ y_1, y_2, \ldots, y_n \geq 0. \end{array} \quad \ldots(4)$$

Changing the above problem to maximization and multiplying each constraint by -1, we get

$$\text{Max. } Z_y' = c_1 y_1 + c_2 y_2 + \dots + c_n y_n, \; (-Z_y = Z_y' \text{ say})$$

subject to

$$\begin{aligned} a_{11} y_1 + a_{12} y_2 + \dots + a_{1n} y_n &\le b_1 \\ a_{21} y_1 + a_{22} y_2 + \dots + a_{2n} y_n &\le b_2 \\ \dots \quad \dots \quad \dots \\ \dots \quad \dots \quad \dots \\ a_{m1} y_1 + a_{m2} y_2 + \dots + a_{mn} y_n &\le b_m, \\ y_1, y_2, \dots, y_n &\ge 0. \end{aligned} \qquad \dots(5)$$

Which is identical to the given linear programming problem (primal problem).

Hence, the dual of the dual is the primal.

Theorem 4:

If x is any feasible solution to the primal problem Max $Z_P = c\,x$, *subject to* $A\,x \le b$, $x \ge 0$ *and w is any feasible solution to the dual problem*

$$Min\ Z_D = b'w$$

subject to $A'w \ge c'$, $w \ge 0$,

then $c\,x \le b'w$ *i.e.,* $Z_P \le Z_D$.

Proof:

Consider the primal problem

$$\text{Max. } Z_P = c\,x$$

$$\text{subject to } A\,x \le b \qquad \dots(1)$$

$$x \ge 0.$$

Let $x = (x_1, x_2, \dots, x_n)$ be any feasible solution to (1).

The dual of the above primal is

$$\text{Min. } Z_D = b'\,w$$

$$\text{subject to } A'\,w \ge c', \; w \ge 0. \qquad \dots(2)$$

Let $w = (w_1, w_2, \dots w_m)$ be any feasible solution to the dual (2).

Now w (m-component column vector) is the feasible solution of the dual and $A\,x \le b$ are the constraints of the primal (1). If we multiply both sides of $A\,x \le b$ with w', the sign of the inequality remains unchanged.

$$\therefore \quad w'(A\,x) \le w'\,b$$

$$\Rightarrow \quad (A'\,w)'\,x \le (b'\,w)', \qquad \dots(3)$$

Similarly, x (n-component column vector) is the feasible solution of the primal (1) and $A'w \geq c'$ denotes the constraints of the dual (2) so we can write

$$x' (A' w) \geq x' c'$$

$$\Rightarrow \quad x' (w' A)' \geq (c x)'$$

$$\Rightarrow \quad [(w' A)x]' \geq (c x)'$$

$$\Rightarrow \quad (w' A) x \geq c x$$

$$\Rightarrow \quad (A' w)' x \geq c x. \qquad ...(4)$$

From relations (3) and (4) we have

$$c x \geq (A' w)' x \leq (b' w)'$$

$$\Rightarrow \quad c x \leq (b' w)' \Rightarrow c x \leq b' w$$

$$\Rightarrow \quad Z_P \leq Z_D.$$

Theorem 5:

If $\hat{x}$ is feasible solution to the primal problem Max. $Z_P = c x$ subject to $A x \leq b$, $x \geq 0$ and $\hat{w}$ is a feasible solution to its dual

$$Min.\ Z_D = b' w \text{ subject to } A' w \geq c',\ w \geq 0$$

such that $c \hat{x} = b' \hat{w}$, then x is the optimal solution of the primal and $\hat{w}$ is the optimal solution of the dual problem.

or

The necessary and sufficient condition for any linear programming problem and its dual to have optimal solution is that both have feasible solution.

Proof:

Consider the primal problem

$$\text{Max. } Z_P = c x$$

subject to $A x \leq b$, $x \geq 0$. ...(1)

Let x be any feasible solution to (1).

The dual of the problem (1) is

$$\text{Min. } Z_D = b' w \qquad ...(2)$$

subject to $A' w \geq c'$, $w \geq 0$

Let $\hat{w}$ be any feasible solution to the dual (2).

Proceeding as in theorem (2) above, we have

$$c\,x \le b'\,\hat{w}$$

$$\Rightarrow \quad c\,x \le c\,\hat{x}, \text{ since } c\,\hat{x} = b'\,\hat{w}$$

Value of the objective function of the primal problem at the feasible solution $\hat{x}$ is greater than its value at any other feasible solution $\hat{x}$.

Hence $\hat{x}$ is the optimal solution of the primal problem (for maximization problem).

Again suppose w is any feasible solution of the dual problem (2). x is the given feasible solution of the primal problem (1),

Proceeding as in theorem (2), we have

$$c\,x \le b'\,w$$

$$\Rightarrow \quad b'\,\hat{w} \le b'\,w \text{ since } c\,x = b'\,\hat{w}$$

The value of the objective function of the dual problem at the given feasible solution $\hat{w}$ is less than its value at any other feasible solution $\hat{w}$.

Hence $\hat{w}$ is the optimal solution of the dual problem (for minimization problem).

Theorem 6:

Prove that dual of a dual is primal problem.

Proof:

Let primal problem is

Max. $Z = c\,x$

s.t. $A\,x \le b$ ...(1)

$x \ge 0$

Dual is Mini. $Z_{D1} = b'\,w$

s.t. $A'\,w \ge c'$ ...(2)

and $w \ge 0$

Thus dual (2) can be written as

Max $(-Z_{D1}) = -\,b'\,w$

$= d\,w$ [Put $-\,b' = d$]

s.t. $-\,A'\,w \le -\,c'$ ...(3)

$w \ge 0$

Dual of the 3rd constraint is

Mini $(-Z_{D1}) = (-c')'\,y$

s.t. $(-A')'\,y \ge (-b')'$ ...(4)

$y \geq 0$

The dual (4) can be written as

Max. $Z_{D2} = c\,y$

s.t. $A\,y \leq b$

$y \geq 0$

which is similar to (1) *i.e.*, the primal.

Theorem 7:

If Max. $Z_p = c\,x$

s.t. $A\,x \leq b$ *is primal*

$x \geq 0$

Mini. $Z_D = b'\,w$

s.t. $A'\,w \geq c'$ *is dual*

$w \geq 0$

Prove that $D_P \leq Z_D$ $\forall$ *x is feasible solution of primal.*

$\forall$ *w is feasible solution of dual.*

Proof:

Primal Max. $Z_P = c\,x$

s.t. $A\,x \leq b$ ...(1)

$x \geq 0$

Dual Mini. $Z_D = b'\,w$

s.t. $A'\,w \geq c'$ ...(2)

$w \geq$

x and w be any feasible solution to the primal (1) and dual problem (2) respectively.

If we multiply the w' a row matrix in the constraint of primal

$$w'\,A\,x \leq w'\,b \qquad ...(3)$$

and multiply the x' a column matrix in the constraint of dual

$$x'\,A'\,w \leq x'\,c' \qquad ...(4)$$

equation (4) can be written as

$$(A\,x)'\,w \geq (c\,x)'$$

Taking transpose both side

[(A x)' w]' A' x' = (x A)' [(c x)]' Matrix problem A' x' = (x A)'

$(A\ x)'' = A\ x$

$w'\ (A\ x)'' \geq (c\ x)''$

$w'\ A\ x \geq c\ x$ ...(5)

from (1) and (5)

$w'\ A\ x \geq Z_P$ ...(6)

from (3) and (6)

$w'\ b \geq w'\ A\ x \geq Z_P$

$(b'\ w)' \geq Z_P$

(Since $(b'\ w)' = b'\ w$, as b' w is a scalar value, not a matrix)

So $b'\ w \geq Z_P$ ...(7)

from (2) and (7)

$Z_D \geq Z_P$.

Theorem 8:

If $\overline{x}$ and $\overline{w}$ are feasible solution of primal & usual respectively such that $c\ \overline{x} = b'\ \overline{w}$ (i.e., $Z_P = Z_D$) then x and w are optimum solution of the respective problem respectively i.e., $\overline{x}$ is max and $\overline{w}$ is mini problem

Max $Z_P = c\ x$

s.t. $A\ x \leq b,\ x \geq 0$

and its corresponding dual

Mini. $Z_D = b'\ w$

$A'\ w \geq c'\ w \geq 0.$

Or Alternative Statement

The necessary and sufficient condition for any L.P.P. and its dual to have optimal solution is that both have feasible solution.

Proof:

Let $\overline{x}$ is any feasible solution to the primal

Max. $Z_P = c\ \overline{x}$, ...(1)

s.t. $A\ x \leq b,\ x \geq 0$

and w is any feasible solution of dual

Mini. $Z_D = b'\ w$

s.t. $A'\ w \geq c',\ w \geq 0$ ...(2)

then from theorem (2), we have

$c\,\overline{x} \leq b'\,w$ $(Z_P \leq Z_D)$

$b'\,\overline{w} = c\,x \leq b'\,w$ (as $c\,\overline{x} = b'\,w$ given)

i.e., $b'\,\overline{w} \leq b'\,w$

$\Rightarrow \overline{w}$ is mini solution of dual let $\overline{w}$ is given feasible solution of dual x is any feasible solution of primal

$c\,x \leq b'\,\overline{w} = c\,\overline{x}$

$\Rightarrow$ $c\,x \leq c\,\overline{x}$

$\Rightarrow$ $c\,\overline{x}$ is max value of the objective function

$\Rightarrow$ $\overline{x}$ is maximum solution of the primal problem.

Theorem 9:

If the L.P.P. mini. $Z_P = c\,x$ & & $A\,v \geq b$ has finites optimal solution than the dual max. $Z = b'\,w$ & $A'\,w \leq c'$ also has a finite maxi solution and if dual has a finite optimal solution then so as primal.

Proof:

Let the primal is

Mini $Z = c\,x$

$a_{11}\,x_1 + a_{12}\,x_2 + \ldots + a_{1n}\,x_n \geq b_1$

$a_{m1}\,x_1 + a_{m2}\,x_2 + \ldots + a_{mn}\,x_n \geq b_m$

changing in equation

$a_{11}\,x_1 + a_{12}\,x_2 \ldots + a_{1n}\,x_n - x_{1s} = b_1$

$a_{m1}\,x_1 + a_{m2}\,x_2 + \ldots + a_{mn}\,x_n - x_{ms} = b_m$

The L.P.P. can be written as

$A\,x - I\,X_s = b$

Let x_B is a basic feasible solution which is optimal for the primal corresponding to base $B = (\alpha_1, \alpha_2, \ldots \alpha_m)$

$\Rightarrow x_B$ minimum solution of primal

$\Rightarrow$for this solution all $Z_j - c_j \leq 0$

$\Rightarrow c_B\,B^{-1}\,\alpha_j - c_j \leq 0$ for all j either original or for surplus variables

$\Rightarrow c_B\,B^{-1}\,\alpha_1 \leq c_1$

$c_B\,B^{-1}\,\alpha_2 \leq c_2$

$c_B\ B^{-1}\ \alpha_n \leq c_n$ *i.e.,* for original variable

$\Rightarrow (c_B\ b^{-1})\ (\alpha, \alpha, \ldots, \alpha_n \leq (c_1, c_2 - c_n)$

$\Rightarrow w_B'\ A \leq c$

Taking transport let $(c_B\ B^{-1} = w_B')$

$\Rightarrow A'\ w_B \leq c'$

$\Rightarrow w_B$ satisfy $A'\ w \leq c'$ *i.e.,* of dual

$\Rightarrow w_B$ is a dual solution.

I[st] we prove w_B is feasible solution IInd we prove w_B is optimum solution

$w_B' = c_B\ (B^{-1})$

Since $Z_J - c_j \leq 0$ for surplus variables also

$\Rightarrow c_B\ B^{-1}\ \alpha_{js} - 0 \leq 0$

$$\Rightarrow (w_{B1}, w_{B2}, \ldots, w_{Bm}) \begin{pmatrix} 0 \\ 0 \\ 0 \\ 1 \\ -1 \\ 0 \end{pmatrix}_{j^{\text{the}}\text{place}} \leq 0$$

$\Rightarrow - w_{BJ} \leq 0$ for $j = 1, 2, -m$

$\Rightarrow w_{BJ} \geq 0\ j = 1, 2, \ldots, m$

$\Rightarrow w_B = (c_B\ B^{-1})$

is a feasible solution of dual.

Part II.

Since $Z_P = (c_B\ B^{-1})\ b$

$= w_B'\ b$

$= b'\ w_B$

$= Z_D$

$\Rightarrow w_B = (c_B\ B^{-1})^1$ is max solution of dual so it is proved that if L.P.P. Min $Z_D = c\ x$ & $A\ x \geq b$ has an finite optimum solution then the dual max $Z_D = b'\ w$ and $A'\ w \leq c'$ also has a maximum solution.

Theorem 10:

If the primal problem has an unbounded solution, then the dual has either no solution or an unbounded solution.

Proof:

We shall prove this theorem by contradiction we suppose that primal has an unbounded solution, dual has a finite optimal solution. By theorem we have proved that dual of dual is primal and if primal has an optimal feasible solution then dual also has a finite optimal solution. Thus if we consider the dual as primal then its dual must have a finite optimal solution, which is a contradiction hence the dual has no finite optimal solution, and if it has it will be unbounded.

Theorem 11:

If there does not exist any finite optimum solution to the primal (dual) then there does not exist any feasible solution to the dual (primal).

Proof:

Suppose the primal problem does not possess any finite optimum solution. Without loss of generality, it can be assumed that there exists a feasible solution to the primal. Now, if we assume that the dual problem possesses a feasible solution by the primal problem must posses a finite optimum solution. This constricted the hypothesis, hence the dual problem does not possess a feasible solution.

Theorem 12:

(Basic Dual problem Theorem) : If x_0 is an optimum solution to the primal, then there exists a feasible w_0 to the dual such that

$$c\,x_0 = b'\,w_0,$$

where b' is the transpose of b.

Proof:

Consider the following primal and dual problems

$$\text{Max. } Z_P = c\,x$$

subject to $A\,x \leq b,\ x \geq 0$ ...(1)

and $\text{Min. } Z_D = b'\,w$

subject to $A'\,w \geq c',\ w \geq 0.$ ...(2)

Suppose x_0 is an optimum solution to the primal. Then to solve the primal problem by simplex method, introduce slack variables to each of the constraints.

Now (1) can be written as

$$\text{Max } Z_P = c\,x$$

subject to $A\,x + Ix_s = b,$

where $x_s \in R^m$ represents the vector of slack variables and I is the associated m × n identity matrix.

Suppose $x_0 = (x_B, 0)$ is an optimum solution to the problem where x_B denotes the optimum basic feasible solution.

We know $x_B = B^{-1} b$, B is the optimal basis of A.

∴ the optimal primal objective function is

$$Z = c\, x_0 = c_B\, x_B,$$

c_B is the row vector containing the prices of the basic variable.

We know

$$Z_j - c_j = c_B\, Y_j - c_j$$

$$= \begin{cases} c_B B^{-1}\alpha_j - c_j, \forall \alpha_j \in A \\ c_B B^{-1} e_j - 0, \forall e_j \in I \end{cases}.$$

Since x_0 is an optimal solution, so we have

$$Z_j - c_j \geq \forall j$$

$$\Rightarrow c_B\, B^{-1}\, a_j \geq c_j,\ c_B\, B^{-1}\, e_j > 0,\ \forall_j$$

$$\Rightarrow c_B\, B^{-1}\, A \geq c,\ c_B\, B^{-1} \geq 0 \text{ (in matrix form)}$$

$$\Rightarrow A'\, B^{-1}\, c_B' \geq c',\ B^{-1}\, c_B \geq 0$$

$$\Rightarrow A'\, w_0 \geq c';\ w_0 \geq 0, \text{ taking } B^{-1}\, c_B = w_0 \in R^m$$

$\Rightarrow w_0$ is a feasible solution of the dual problem. Also corresponding dual objective function is

$$b'\, w_0 = b_0'\, w = c_B\, B^{-1}\, b = c_B\, x_B = c\, x_0$$

Hence corresponding to a given optimal solution x_0 of the primal there exists a feasible solution w_0 of the dual such that

$$c\, x_0 = b'\, w_0.$$

Note: Similarly, we can prove the following theorem:

If w_0 is an optimal solution to the dual, then there exists a feasible solution x_0 to the primal such that $c\, x_0 = b'\, w_0$.

Theorem 13:

(i) If either the primal or the dual problem has a finite optimal solution then the other problem so has a finite optimal solution and the optimal values of the objective function in both the problems are the same.

(ii) If primal (dual) problem has an unbounded optimum solution, the other problem has either no solution at all or an unbounded solution.

Proof:

Consider the primal and the dual problems

Primal problem

$$\text{Max. } Z_P = c\,x$$

subject $A\,x \leq b, x \geq 0$ ...(i)

Dual problem.

$$\text{Min. } Z_D = b'\,w$$

subject to $A'\,w \geq c'$ ...(ii)

$w \geq 0$

To prove the theorem we shall construct an optimal solution to the dual form a given optimal solution of the primal.

Let us assume that the primal has a finite optimal feasible solution x_B.

To solve the primal (i) by simplex method, introduce slack variables to each of the constraints.

Now (i) can be written as

$$\text{Max. } Z_P = c\,x$$

subject $A\,x + Ix_s = b$ ...(iii)

$x \geq 0,\ x_s \geq 0,$

where $x_s \in R^m$ represents the vector of slack variables and I is the associated $m \times n$ identity matrix. B is the basis matrix and suppose c_B is the m-component row vector containing the prices of the basic variables.

Now x_B is the optimal solution to the primal so we have

$c_j - Z_j \leq 0, \forall_j.$

We know $Z_j = c_B\,Y_j = c_B\,B^{-1}\,\alpha_j$

$c_j - c_B\,B^{-1}\,\alpha_j \leq 0, \forall a_j$

or $c_B\,B^{-1}\,\alpha_j \geq c_j$ ...(iv)

$\Rightarrow c_B\,B^{-1}\,(\alpha_1, \alpha_2, \ldots, \alpha_n) \geq (c_1, c_2, \ldots, c_n)$

$\Rightarrow c_B\,B^{-1}\,A \geq c$ ...(v)

Taking $(\hat{w})' = c_B\,B^{-1}$, where $\hat{w} = (w_1, w_2, \ldots, w_m)$.

Then from (v), we have

$(w)'\,A \geq c \Rightarrow [(w)'\,A]' \geq c'$

$\Rightarrow A'\,(\hat{w}) \geq c'$

$\Rightarrow \hat{w}$ is the solution of the dual (ii).

Again considering the relation (iv) with a_j corresponding to slack variables, we have

$$c_B B^{-1} (e_j) \geq 0, j = 1, 2, \ldots, m; c_j = 0.$$

$$\Rightarrow (\hat{w})' e_j \geq 0 \Rightarrow (\hat{w})' \geq 0$$

$$\Rightarrow \hat{w} \geq 0$$

$\Rightarrow$w is the feasible solution of the dual (ii).

Now it remains to show that w is an optimal solution to the dual (ii).

We have

$$Z_D = b' \hat{w} = [(\hat{w})' b]' = (\hat{w})' b$$

$$= (c_B B^{-1}) b = c_B (B^{-1} b)$$

$$= c_B x_B = Z_P.$$

Since w and x_B are the feasible solution of the dual (ii) and the primal (1) respectively and

$$Z_D = Z_P$$

therefore $\hat{w}$ is the optimal solution of the dual (ii).

(ii) We shall prove it by contradiction. Suppose, when the primal problem has an unbounded solution, the dual problem has a finite optimal solution.

We have proved in theorem 1 that dual of the dual is the original primal. Also part (i) of this theorem we have proved that if a primal has a finite optimal solution, the dual also has finite optimal solution.

Thus, if we consider the dual as the primal, then its (which is the original primal) must have a finite optimal solution, which contradicts the hypothesis. Hence the dual has no finite optimal solution. If the primal problem has an unbounded solution, either the dual has no solution or an unbounded solution.

Similarly, we can prove that if the dual has an unbounded solution, the primal has no solution or an unbounded solution.

Theorem 14:

If any of the constraints in the primal is a perfect equality, the corresponding dual variable is unrestricted in sign.

Proof:

Suppose in the given primal the kth constraint is an equality. Writing the primal in the standard primal form, we have

$$\text{Max. } Z = c_1 x_1 + c_2 x_2 + \ldots + c_n x_n$$

subject to

$$a_{11} x_1 + a_{12} x_2 + \ldots + a_{1n} x_n \le b_1$$
$$a_{21} x_1 + a_{22} x_2 + \ldots + a_{2n} x_n \le b_2$$
$$\ldots \quad \ldots \quad \ldots$$
$$\ldots \quad \ldots \quad \ldots$$
$$a_{k1} x_1 + a_{k2} x_2 + \ldots + a_{kn} x_n \le b_k$$
$$-a_{k1} x_1 - a_{k2} x_2 - \ldots - a_{kn} x_n \le -b_k$$
$$\ldots \quad \ldots \quad \ldots$$
$$\ldots \quad \ldots \quad \ldots$$
$$a_{m1} x_1 + a_{m2} x_2 + \ldots + a_{mn} x_n \le b_m,$$
$$x_1, x_2, \ldots, x_n \ge 0.$$

The dual of the above primal is

$$\text{Min } Z_D = b_1 w_1 + b_2 w_2 + \ldots + b_k (w_k' \;\; w_k'') + \ldots + b_m w_m$$

subject to

$$a_{11} w_1 + a_{21} w_2 + \ldots + a_{k1} (w_k' - w_k'') + \ldots + a_{m1} w_m \ge c_1$$
$$a_{12} w_1 + a_{22} w_2 + \ldots + a_{k2} (w_k' - w_k'') + \ldots + a_{m2} w_m \ge c_2$$
$$\ldots \quad \ldots \quad \ldots$$
$$\ldots \quad \ldots \quad \ldots$$
$$a_{1n} w_1 + a_{2n} w_2 + \ldots + a_{kn} (w_k' - w_k'') + \ldots + a_{mn} w_m \ge c_n$$
$$w_1, w_2, \ldots, w_k', w_k'', \ldots, w_m \ge 0.$$

Substituting $w_k = w_k' - w_k''$ in the above dual, we get

$$\text{Min. } Z_D = b_1 w_1 + b_2 w_2 + \ldots + b_k w_k + \ldots + b_m w_m$$

subject to

$$a_{11} w_1 + a_{21} w_2 + \ldots + a_{k1} w_k + \ldots + a_{m1} w_m \ge c_1$$
$$a_{12} w_1 + a_{22} w_2 + \ldots + a_{k2} w_k + \ldots + a_{m2} w_m \ge c_2$$
$$\ldots \quad \ldots \quad \ldots$$
$$\ldots \quad \ldots \quad \ldots$$
$$a_{1n} w_1 + a_{2n} w_2 + \ldots + a_{kn} w_k + \ldots + a_{mn} w_m \ge c_n$$
$$w_1, w_2, \ldots, w_k - 1, w_k + 1, \ldots, w_m \ge 0.$$

wk is unrestricted in sign because

$w_k > 0$ if $w_k' > w_k''$ and $w_k < 0$ if $w_k' < w_k''$.

Theorem 15:

There exist a bounded (finite) optimum solution to a linear programming problem if and only if there exists a feasible solution to both primal and its dual.

Proof:

Suppose Max. $Z = c\,x$

subject to $A\,x \leq b,\ x \geq 0$

and Min. $Z_D = b'\,w$

subject to $A'\,w \geq c',\ w \geq 0$

are the primal and dual problems respectively.

Again suppose there exists an optimum feasible solution to the primal problem. Then by fundamental theorem of Dual problem the dual problem has atleast one feasible solution. Conversely, let us assume that both primal and the dual possess feasible solution x and w respectively. Then $c\hat{x}$ and $b'\,\hat{w}$ are both finite and $c\,\hat{x} \leq b'\,\hat{w}$ *i.e.,* $b'\,\hat{w}$ acts as an upper bound on $c\,\hat{x}$ although not necessarily the least upper bound. Hence the primal must have finite optimum solution.

Theorem 16:

If any variable of the primal is unrestricted in sign, the corresponding constraint in the dual will be a strict equality.

Proof:

Consider the primal in which the kth variable is unrestricted in sign.

$$\text{Max. } Z = c_1 x_1 + c_2 x_2 + .. + c_k x_k + ... + c_n x_n$$

subject to

$$a_{11} x_1 + a_{12} x_2 + ... + a_{1k} x_k + ... + a_{1n} x_n \leq b_1$$

$$a_{21} x_1 + a_{22} x_2 + ... + a_{2k} x_k + ... + a_{2n} x_n \leq b_2$$

... ...

... ...

$$a_{m1} x_1 + a_{m2} x_2 + ... + a_{mk} x_k + ... + a_{mn} x_n \leq b_m$$

$$x_1, x_2, ..., x_k - 1, x_{k+1} ..., x_n \geq 0,$$

x_k is unrestricted in sign.

Substituting $x_k = x_k' - x_k''$, $x_k' \geq 0$ in the given primal, it changes to

$$\text{Max. } Z = c_1 x_1 + c_2 x_2 + \ldots + c_k (x_k' - x_k'') + .. + c_n x_n$$

subject to

$$a_{11} x_1 + a_{12} x_2 + \ldots + a_{1k} (x_k' - x_k'') + \ldots + a_{1n} x_n \leq b_1$$
$$a_{21} x_1 + a_{22} x_2 + \ldots + a_{2k} (x_k' - x_k'') + \ldots + a_{2n} x_n \leq b_2$$
$$\ldots \quad \ldots \quad \ldots$$
$$\ldots \quad \ldots \quad \ldots$$
$$a_{m1} x_1 + a_{m2} x_2 + \ldots + a_{mk} (x_k' - x_k'') + \ldots + a_{mn} x_n \leq b_m,$$
$$x_1, x_2, \ldots, x_{k-1}, x_k', x_k'', x_{k+1}, \ldots, x_n \geq 0.$$

Writing the dual of the above primal, we have

$$\text{Min. } Z_D = b_1 w_1 + b_2 w_2 + .. + b_m w_m,$$

subject to

$$a_{11} w_1 + a_{21} w_2 + \ldots + a_{m1} w_m \geq c_1$$
$$a_{12} w_1 + a_{22} w_2 + \ldots + a_{m2} w_m \geq c_2$$
$$\ldots \quad \ldots \quad \ldots$$
$$\ldots \quad \ldots \quad \ldots$$
$$a_{1k} w_1 + a_{2k} w_2 + \ldots + a_{mk} w_m \geq c_k$$
$$- a_{1k} w_1 - a_{2k} w_2 - \ldots - a_{mk} w_m \geq - c_k$$
$$\ldots \quad \ldots \quad \ldots$$
$$\ldots \quad \ldots \quad \ldots$$
$$a_{1n} w_1 + a_{2n} w_2 + \ldots + a_{mn} w_m \geq c_n,$$
$$w_1, w_2, \ldots, w_m \geq 0.$$

The two constraints

$$a_{1k} w_1 + a_{2k} w_2 + \ldots + a_{mk} w_m \geq c_k$$

and $$- a_{1k} w_1 - a_{2k} w_2 - \ldots - a_{mk} w_m \geq - c_k$$

are equivalent to the single equation

$$a_{1k} w_1 + a_{2k} w_2 + \ldots + a_{mk} w_m = c_k.$$

Hence if the kth variable of the primal is unrestricted, the kth constraint in the dual is an equality.

Theorem 17:

If there does not exist feasible solution to the dual (primal) but there exists at least one to the primal (dual), then there dies not exist any finite optimum solution to the primal (dual).

Proof:

Consider the primal problem

$$\text{Max. } Z = c\,x,$$

subject to $A\,x \leq b,\ x \geq 0$.

The corresponding dual problem is

$$\text{Min. } Z_D = b'\,w$$

subject to $A'\,w \geq c',\ w \geq 0$.

Suppose there does not exist any feasible solution to the dual there does one to the primal, say $\hat{x}$. Then $c\,\hat{x}$ is the value of the primal objective function.

If we suppose that $\hat{x}$ is an optimum solution to the primal then by fundamental theorem of Dual problem there must exist a feasible solution to the dual which contradicts the hypothesis.

Therefore no feasible solution to the primal can be optimal.

Similarly, we can start with the primal.

SOME USEFUL ASPECTS OF DUAL PROBLEM

In some cases it is easier to solve linear programming problems through their duals. If the number of original variables in the primal problem is considerably less than the number of slack and surplus variables, we must opt to solve the problem through its dual.

Dual problem plays an important role not only in linear programming but also in physics, economics, engineering etc.

In physics, we use it in parallel circuit and series circuit theory.

In Economics it is used in the formulation of input and output systems.

In game theory, we use it to find the optimal strategies of the player B when he minimizes his losses. Then, using Dual problem, we can change the player A's problem into player B's problem and vice-versa.

ADVANTAGES OF DUALITY

1. It is advantageous to solve the dual of a primal having less number of constraints, because the number of constraints usually equals the number of iterations required to solve the problem.
2. It avoids the necessity for adding surplus or artificial variables and solves the problem quickly (the technique is known as the primal-dual

method). In economics, duality is useful in the formulation of the input and output systems. It is also useful in physics, engineering, mathematics, etc.

3. The dual variables provide an important economic interpretation of the final solution of an LP problem.
4. It is quite useful when investigating changes in the parameters of an LP problem (the technique is known as the sensitivity analysis).
5. Duality is used to solve an LP problem by the simplex method in which the initial solution is infeasible (the technique is known as the dual simplex method).

COMPLEMENTARY SLACKNESS THEOREM

For the optimal feasible solution of the primal and the dual systems:

(i) If the inequality occurs in the ith relation of either system (the corresponding slack or surplus variable if positive), then the ith variable of its dual is zero.

(ii) If the ith variable if positive in either system, the ith relation of its dual holds as a strict equality (i.e., the corresponding slack or surplus variable $w_{m+j} = 0$.)

Proof:

The objective functions of the primal and dual problems in explicit form can be written as

$$\text{Max. } Z_P = c_1 x_1 + c_2 x_2 + \dots + c_n x_n \text{ (primal)} \quad \dots(1)$$

$$\text{Min. } Z_D = b_1 w_1 + b_2 w_2 + \dots + b_m w_m \text{ (dual)} \quad \dots(2)$$

After introducing the non-negative slack variables $x_{n+1}, x_{n+2}, \dots, x_{n+m}$ the primal constraint equations can be written as

$$\begin{aligned} a_{11} x_1 + a_{12} x_2 + \dots + a_{1n} x_n + x_{n+1} &= b_1 \\ a_{21} x_1 + a_{22} x_2 + \dots + a_{2n} x_n + x_{n+2} &= b_2 \quad \dots(3) \\ \dots \quad \dots \quad \dots & \\ \dots \quad \dots \quad \dots & \\ a_{m1} x_1 + a_{m2} x_2 + \dots + a_{mn} x_n + x_n + m &= b_m \\ x_1, x_2, \dots, x_{n+m} &\geq 0. \end{aligned}$$

Again introducing the no-negative surplus variables $w_{m+1}, w_{m+2}, \dots, w_{m+n}$ the dual constraints equations can be written as

$$a_{11} w_1 + a_{21} w_2 + \dots + a_{m1} w_m - w_{m+1} = c_1$$

$$a_{12} w_1 + a_{22} w_2 + \dots + a_{m2} w_m - w_{m+2} = c_2 \qquad \dots(4)$$

$$\dots \qquad \dots \qquad \dots$$

$$\dots \qquad \dots \qquad \dots$$

$$a_{1n} w_1 + a_{2n} w_2 + a_{mn} w_m - w_{m+n} = c_n$$

$$w_1, w_2, \dots, w_m, w_{m+1}, \dots, w_{m+n} \geq 0.$$

Now, multiplying the equations (3) by $w_1, w_2, \dots, w_m$ respectively and then adding the resulting equations, we have

$$x_1 \sum_{i=1}^{m} a_{i1} w_i + x_2 \sum_{i=1}^{m} a_{i2} w_i + \dots + x_n \sum_{i=1}^{m} a_{in} w_i$$

$$+ x_{n+1} w_1 + x_{n+2} w_2 + \dots + x_{n+m} w_m = \sum_{i=1}^{m} b_i w_i \qquad \dots(5)$$

subtracting (5) from (1), we have

$$\left(c_1 - \sum_{i=1}^{m} a_{i1} w_i\right) x_1 + \left(c_2 - \sum_{i=1}^{m} a_{i2} w_i\right) x_2 + \dots + \left(c_n - \sum_{i=1}^{m} a_{in} w_i\right) x_n$$

$$- w_1 x_{n+1} - w_2 x_{n+2} - \dots - w_m x_{n+m} = \left(Z_P - \sum_{i=1}^{m} b_i w_i\right)$$

$$= Z_P - Z_D \qquad \dots(6)$$

From (4), we have

$$- w_{m+j} = c_j - (a_{1j} w_1 + a_{2j} w_2 + \dots + a_{mj} w_m),\ j = 1, 2, \dots, n$$

$$\text{or} - w_{m+j} = c_j - \sum_{i=1}^{m} a_{ij} w_i \text{ for all } j = 1, 2, \dots, n. \qquad \dots(7)$$

Using (7) in (6), we have

$$- (w_{m+1} x_1 + w_{m+2} x_2 + \dots + w_{m+n} x_n)$$

$$- (w_1 x_{n+1} + w_2 x_{n+2} + \dots + w_m x_{n+m}) = Z_P - Z_D \qquad \dots(8)$$

Now if $x^* = (x_1^*, x_2^*, \dots, x_n^*)$

and $w^* = (w_1^*, w_2^*, \dots, w_m^*)$

be the optimal solution to the primal and the dual problems respectively, then by Dual problem theorem

$$Z_P^* = Z_D^*.$$

Thus for the optimal solution x^* and w^* of the primal and dual problems the corresponding slack and surplus variables are

$x^*_{n+1} \geq 0$, i = 1, 2, ... , m and $w^*_{m+j} \geq 0$, = 1, 2, ... , n.

Form equation (8), we have

$$(w^*_{m+1}\, x_1^* + w^*_{m+2}\, x_2^* + \ldots + w^*_{m+n}\, x_n^*)$$
$$+ (w_1^*\, x^*_{n+1} + w_2^*\, x^*_{n+2} + \ldots + w_m^*\, x^*_{n+m}) = 0 \qquad \ldots(9)$$

Since all the variables in (9) are non-negative so their product terms in (9) are also non-negative. For the validity of relation (9) each term must be individually equal to zero

i.e., $w^*_{m+j}\, x_j^* = 0\ (\forall_j = 1, 2, \ldots, n)$ ...(10)

and $w_i^*\, x^*_{n+i} = 0\ (\forall_i = 1, 2, \ldots, m)$ ...(11)

(i) Form (11), if $x^*_{n+1} > 0$ then we must have $w_i^* = 0$ *i.e.*, if the slack variable in the ith relation of the primal is positive then ith variable of the dual is zero.

Again from (10), if $w^*_{m+j} > 0$ then we must have $x_j^* = 0$ *i.e.*, if the surplus variable in the ith relation of the dual is positive then jth variable of the primal (dual of the dual) is zero.

This proves the part (i) of the theorem.

(ii) Also from (10), if $x_j^* > 0$, then we must have $w^*_{m+j} = 0$ *i.e.*, if the ith variable in the primal is positive then the jth relation in the dual is strictly an equality (as $w^*_{m+j} = 0$).

Again from (11), if

$$w_i^* > 0, \text{ then } x^*_{n+j} = 0$$

i.e., if the ith variable in the dual is positive then the jth relation in the primal (dual of the dual) is strictly an equality (as $x^*_{n+i} = 0$).

This proves the part (ii) of the theorem.

Alternative Statement

A necessary and sufficient condition for any pair of feasible solutions to the primal and dual to be optimal is that

$$w_i\, x_{n+i} = 0,\ i = 1, 2, \ldots, m$$

where x_{n+i} is the slack variable in the primal

and $x_j\, w_{m+j} = 0,\ j = 1, 2, \ldots, n,$

where w_{m+j} is the surplus variables for the dual.

Primal and Dual Correspondence

With the help of the Dual problem theorems, we note the following correspondence between the *primal* and the *dual* problems.

	Primal	*Dual*
1.	Objective function Max. Z_P.	Objective function Min. Z_D.
2.	Requirement vector.	Price vector.
3.	Coefficient matrix A	Transpose of the coefficient matrix, A' or A^T.
4.	Constraint with ≤ sign.	Constraints with ≥ sign.
5.	Relation.	Variable.
6.	i-th inequality.	i-th variable $w_i \geq 0$.
7.	i-th constraint an equality.	i-th variable unrestricted in sign
8.	Variable.	Relation.
9.	i-th variable $x_i > 0$.	i-th relation a strict inequality.
10.	i-th variable x_i unrestricted in sign.	i-th constraint a strict equality.
11.	i-th slack variable positive.	i-th variable zero.
12.	i-th variable zero.	i-th surplus variable positive.
13.	Finite optimal solution.	Finite optimal solution with equal optimal value of objective function.
14.	Unbounded solution	No solution or an unbounded solution.

Rules for Obtaining the Solution to the Dual from the Final Simplex Table of the Primal and Vice-Versa

From the final simplex table of the primal problem we can also read the optimal solution of the dual and vice-versa. For this we observe the following rules:

1. *The optimal value of the primal objective function is equal to the optimal value of the dual objective function.*

   i.e., $\quad Max.\ Z_P = Min.\ Z_D.$

   This has been proved in Dual problem Theorem.

2. *The Δ_j's ($\Delta_j = c_j - Z_j$) with sign changed for the lack (or surplus) variables in the final simplex table for the primal are the values of the corresponding optimal dual variables in the final simplex table for the dual.*

Proof:

In the final simplex table (for primal) corresponding to slack and surplus variables

$$-\Delta_j = -(c_j - Z_j) = -(0 - Z_j)$$
$$= c_B Y_j = c_B B^{-1} \alpha_j$$

where B is the optimal basis and c_B, the corresponding price vector.

But, for i-th slack variable, a_j becomes a unit vector e_i with 1 at the i-th place.

$$\therefore \qquad -\Delta_j = c_B B^{-1} e_i = (w_1, w_2, \dots, w_m)\, e_i = w_i$$

where w_j is i-th dual variable.

3. If either problem has unbounded solution, then the other will have no feasible solution.

Note: The simplex method to the problem having lesser number of constraints. Therefore, first we shall solve the problem (primal or dual) with lesser number of constraints by simplex method.

Example 1:

Write the dual of the L.P. problem

$$\text{Max.} \quad Z = 2x_1 + x_2$$

$$\text{s.t.} \quad x_1 + 2x_2 \le 10$$

$$x_1 + x_2 \le 6$$

$$x_1 - x_2 \le 2$$

$$x_1 - 2x_2 \le 1$$

$$x_1, x_2, x_3 \ge 0.$$

Solution:

In matrix from the given problem can be written as

Max. $Z = (2, 1)\,(x_1, x_2) = c\,x$

$$\text{s.t.} \quad \begin{bmatrix} 1 & 2 \\ 1 & 1 \\ 1 & -1 \\ 1 & -2 \end{bmatrix} \begin{bmatrix} x_1 \\ x_2 \end{bmatrix} \le \begin{bmatrix} 10 \\ 6 \\ 2 \\ 1 \end{bmatrix}$$

or $\quad AX \le b$

The dual of the given problem is

$$\text{Mini. } Z_D = b'\,w$$

$$= [10, 6, 2, 1]\,[w_1, w_2, w_3, w_4]$$

$$Z_D = 10w_1 + 6w_2 + 2w_3 + w_4$$

s.t. $\quad A'\,w \ge c'$

$$\text{or} \begin{bmatrix} 1 & 1 & 1 & 1 \\ 2 & 1 & -1 & -2 \end{bmatrix} \begin{bmatrix} w_1 \\ w_2 \\ w_3 \\ w_4 \end{bmatrix} \ge \begin{bmatrix} 2 \\ 1 \end{bmatrix}$$

$$w_1 + w_2 + w_3 + w_4 \geq 2$$

$$2w_1 + w_2 - w_3 - 2w_4 \geq 1$$

$$w_1, w_2, w_3, w_4 \geq 0$$

Example 2:

Consider the symmetric primal problem

Max. $Z_x = 5x_1 + 9x_2$

subject to $x_1 \leq 6$

$$x_1 + x_2 \leq 13$$

$$x_2 \leq 8,$$

$$x_1, x_2 \geq 0.$$

Solution:

The corresponding dual problem is

Min. $Z_w = 6w_1 + 13w_2 + 8w_3$

subject to $w_1 + w_2 \geq 5$

$$w_2 + w_3 \geq 9,$$

$$w_1, w_2, w_3 \geq 0.$$

Note: The following table a simple and convenient method to remember the primal dual relationship:

$(x_1, x_2, \ldots, x_n)$ Min.

$$\begin{bmatrix} w_1 \\ w_2 \\ \vdots \\ w_m \end{bmatrix} \begin{bmatrix} a_{11} & a_{12} & \cdots & a_{1n} \\ a_{21} & a_{22} & \cdots & a_{2n} \\ \cdots & \cdots & \cdots & \cdots \\ a_{m1} & a_{m2} & \cdots & a_{mn} \end{bmatrix} \leq \begin{bmatrix} b_1 \\ b_2 \\ \vdots \\ b_m \end{bmatrix}$$

Max. $(c_1, c_2, \ldots, c_n)$

Variables then it is always possible to obtain the dual of this system. In fact every equation is equal to two inequalities one with less than and second with greater than inequality and every unrestricted variable is equal to difference of two positive variable. A constraint with $\leq$ sign can be converted into a constraint with $\geq$ sign by multiply by -1.

Example 3:

Write the dual of the problem.

Maxi. $Z = 13x_1 + x_2$

s.t. $12x_1 + 3x_2 \leq 20$

$4x_1 + x_2 \leq 35.$

Solution:

In matrix form the given problem can be written as

Max. $Z = (13, 1)(x_1, x_2)$

$= cx$

s.t. $\begin{bmatrix} 12 & 3 \\ 4 & 1 \end{bmatrix} \begin{bmatrix} x_1 \\ x_2 \end{bmatrix} \leq \begin{bmatrix} 20 \\ 35 \end{bmatrix}$

or $ax \leq b.$

The dual of the given problem is

Mini. $Z_D = b' w$

$= (20, 35)(w_1, w_2)$

Mini. $Z_D = 20w_1 + 35w_2$

s.t. $A' w \geq c'$

$$\begin{bmatrix} 12 & 4 \\ 3 & 1 \end{bmatrix} \begin{bmatrix} w_1 \\ w_2 \end{bmatrix} \geq \begin{bmatrix} 13 \\ 1 \end{bmatrix}$$

$12w_1 + 4w_2 \geq 13$

$3w_1 + w_2 \geq 1$

$w_1, w_2 \geq 0.$

Example 4:

Find the dual of the L.P. problem

Mini $Z = 4x_1 + 6x_2 + 18x_3$

s.t. $x_1 + 3x_3 \geq 3$

$x_2 + 2x_3 \geq 5$

$x_1, x_2, x_3 \geq 0.$

Solution:

In matrix from the given problem can be written as

Mini. $Z = 4, 6, 18)(x_1, x_2, x_3)$

$= c\,x$

s.t. $\begin{bmatrix} 1 & 0 & 3 \\ 0 & 1 & 2 \end{bmatrix} \begin{bmatrix} x_1 \\ x_2 \\ x_3 \end{bmatrix} \geq \begin{bmatrix} 3 \\ 5 \end{bmatrix}$

or $AX \geq b$

The dual of the given problem is

Max. $Z_D = b'w$

$= (3, 5)(w_1, w_2)$

$= 3w_1 + 5w_2$

Max. $Z_D = 3w_1 + 5w_2$

s.t. $A'w \leq c'$

$$\begin{bmatrix} 1 & 0 \\ 0 & 1 \\ 3 & 2 \end{bmatrix} \begin{bmatrix} w_1 \\ w_2 \end{bmatrix} \leq \begin{bmatrix} 4 \\ 6 \\ 18 \end{bmatrix}$$

or $1.w_1 + 0w_2 \leq 4$

$0w_1 + 1w_2 \leq 6$

$3w_1 + 2w_2 \leq 18$

$w_1, w_2 \geq 0.$

Example 5:

The dual of L.P. problem.

Max. $Z = 2x_1 + 2x_2$

s.t. $2x_1 - 3x_2 \leq 3$

$4x_1 + x_2 \leq -4$

$x_1, x_2 \geq 0.$

Solution:

In matrix from the given problem can be written as

Max. $Z = (2, 2)(x_1, x_2)$

$= c\,x$

s.t. $$\begin{bmatrix} 2 & -3 \\ 4 & 1 \end{bmatrix} \begin{bmatrix} x_1 \\ x_2 \end{bmatrix} \leq \begin{bmatrix} 3 \\ -4 \end{bmatrix}$$

or $A\,X \leq b$

The dual of the given problem is

Mini $Z_D = b'w$

$= (3, -4)(w_1, w_2)$

$Z_D = 3w_1 - 4w_2$

s.t. $\quad A' w \geq c'$

$$\begin{bmatrix} 2 & 4 \\ -3 & 1 \end{bmatrix} \begin{bmatrix} w_1 \\ w_2 \end{bmatrix} \geq \begin{bmatrix} 2 \\ 2 \end{bmatrix}$$

$2w_1 + 4w_2 \geq 2$

$-3w_1 + w_2 \geq 2$

$w_1, w_2 \geq 0$

Example 6:

Find the dual of the following L.P.P.

Max. $\quad Z = 6x_1 + 5x_2 + 10x_3$

s.t. $\quad 4x_1 + 5x_2 + 7x_3 \leq 5$

$3x_1 + 0x_2 + 7x_3 \leq 10$

$2x_1 + x_2 + 8x_3 \leq 20$

$0x_1 + 2x_2 + 9x_3 \geq 5$

$x_1, x_3 \geq 0$

x_2 an restricted in sign.

Solution:

First of all we shall write the given problem in standard form as follows.

(i) Since it is a maximization problem, all the constraints must contain the sign $\leq$.

Therefore we multiply the fourth constraints by -1 we get

$$-2x_2 - 9x_3 \leq -5.$$

(ii) The variable x_2 is unrestricted in sign

$\therefore$ we write $x_2 = x_2 - x_2''$ where $x_2', x_2'' \geq 0$

So first constraint is equivalent to

$$4x_1 + 5(x_2' - x_2'') + 7x_3 \leq 5$$

The third constraint can be written as

$$2x_1 + (x_2' - x_2'') + 8x_3 \leq 20$$

and fourth constraint is equivalent to

$$2(x_2' - x_2'') + 9x_3 \geq 5$$

or $\quad -2x_2' + 2x_2'' - 9x_3 \leq -5$

Thus the given problem is standard primal form is

$$\text{Max. } Z = 6x_1 + 5(x_2' - x_2'') + 10x_3$$
$$= (6, 5, -5, 10)\,[x_1, x_2', x_2'', x_3]$$
$$= c\,x$$

s.t.
$$4x_1 + 5(x_2' - x_2'') + 7x_3 \le 5$$
$$3x_1 + 7x_3 \le 10$$
$$2x_1 + x_2' - x_2'' + 8x_3 \le 20$$
$$-2x_2' + 2x_2'' - 9x_3 \le -5$$

or
$$\begin{bmatrix} 4 & 5 & -5 & 7 \\ 3 & 0 & 0 & 7 \\ 2 & 1 & -1 & 8 \\ 0 & -2 & 2 & -9 \end{bmatrix} \begin{bmatrix} x_1 \\ x_2' \\ x_2'' \\ x_3 \end{bmatrix} \le \begin{bmatrix} 5 \\ 10 \\ 20 \\ -5 \end{bmatrix}$$

$$A\,x \le b$$

and $x_1, x_2', x_2'', x_3 \ge 0$

The dual of the given problem is

$$\text{Mini. } Z_D = b'\,w$$
$$= [5, 10, 20, -5]\,[w_1, w_2, w_3, w_4]$$
$$Z_D = 5w_1 + 10w_2 + 20w_3 - 5w_4$$

s.t. $A'\,w \le c'$

$$\begin{bmatrix} 4 & 3 & 2 & 0 \\ 5 & 0 & 1 & -2 \\ -5 & 0 & -1 & 2 \\ 7 & 7 & 8 & -9 \end{bmatrix} \begin{bmatrix} w_1 \\ w_2 \\ w_3 \\ w_4 \end{bmatrix} \le \begin{bmatrix} 6 \\ 5 \\ -5 \\ 10 \end{bmatrix}$$

$$4w_1 + 3w_2 + 2w_3 + 0w_4 \ge 6$$
$$5w_1 + 0w_2 + 1w_3 - 2w_4 \ge 5$$
$$-5w_1 + 0w_2 - 1w_3 + 2w_4 \ge -5$$
$$7w_1 + 7w_2 + 8w_3 - 9w_4 \ge 10$$
$$w_1, w_2, w_4 \ge 0$$

and w_3 is unrestricted.

Example 7:

Find the dual of following L.P.P.

Mini. $Z = x_1 + x_2 + x_3$

s.t. $x_1 - 3x_2 + 4x_3 = 5$

$$x_1 - 2x_2 \leq 3$$
$$2x_2 - x_3 \geq 4$$
$$x_1, x_2 \geq 0 \text{ , } x_3 \text{ is unrestricted in sign.}$$

Solution:

First of all we shall write the given problem in standard form as follows:

(i) It is a minimization problem, all the constraints must contain the sign ≥.

Therefore, we multiply the second constraints by –1 we get

$$-x_1 + 2x_2 \geq -3.$$

(ii) The variable x_3 is unrestricted in sign

∴ we write $x_3 = x_3' - x_3''$ where $x_3', x_3'' \geq 0$

So first constraint is equivalent to

$$x_1 - 3x_2 + 4(x_3' - x_3'') \geq 5 - \quad \text{(i)}$$
$$x_1 - 3x_2 + 4(x_3' - x_3'') \leq 5 - \quad \text{(ii)}$$

Multiply above (ii) by –1

$$-x_1 + 3x_2 - 4(x_3' - x_3'') \geq -5$$

Thus the given problem is standard primal form is

Mini. $Z = x_1 + x_2 + x_3' - x_3''$

$= (1, 1, 1, -1)\ [x_1, x_2, x_3', x_3'']$

$= c\,x$

s.t.
$$x_1 - 3x_2 + 4x_3' - 4x_3'' \geq 5$$
$$-x_1 + 3x_2 - 4x_3' + 4x_3'' \geq -5$$
$$-x_1 + 2x_2 + 0x_3' - 0x_3'' \geq -3$$
$$0x_1 + 2x_2 - x_3' + x_3'' \geq 4$$

or
$$\begin{bmatrix} 1 & -3 & 4 & -4 \\ -1 & 3 & -4 & 4 \\ -1 & 2 & 0 & 0 \\ 0 & 2 & -1 & 1 \end{bmatrix} \begin{bmatrix} x_1 \\ x_2 \\ x_3 \\ x_3 \end{bmatrix} \geq \begin{bmatrix} 5 \\ -5 \\ -3 \\ 4 \end{bmatrix}$$

or $A\,x \geq b$

and $x_1, x_2, x_3', x_3'' \geq 0$

∴ The dual of the given problem is

Max. $Z_D = b'\,w$

$= (5, -5, -3, 4)\ (w_1', w_1'', w_2, w_3)$

$= 5w_1' - 5w_1'' - 3w_2 + 4w_3$

s.t. $A'w \geq c'$

$$\begin{bmatrix} 1 & -1 & -1 & 0 \\ -3 & 3 & 2 & 2 \\ 4 & -4 & 0 & -1 \\ -4 & 4 & 0 & 1 \end{bmatrix} \begin{bmatrix} w_1' \\ w_1'' \\ w_2 \\ w_3 \end{bmatrix} \leq \begin{bmatrix} 1 \\ 1 \\ 1 \\ -1 \end{bmatrix}$$

or $w_1' - w_1'' - w_2 + 0.w_3 \leq 1$

$-3w_1' + 3w_1'' + 2w_2 + 2w_3 \leq 1$

$4w_1' - 4w_1'' + 0.w_2 - w_3 \leq 1$

$-4w_1' + 4w_1'' + 0.w_2 + w_3 \leq -1$

and $w_1', w_1'', w_2, w_3 \geq 0$

writing $w_1' - w_1'' = w_1$, the dual problem is also written as

Max. $Z_D = 5w_1 - 3w_2 - 4x_3$

s.t. $w_1 - w_2 \leq 1$

$-3w_1 + 2w_2 + 2w_3 \leq 1$

$4w_1 - w_3 = 1$

$w_2, w_3 \leq 0$, w_1 unrestricted in sign.

Example 8:

Apply the principle of Dual problem to solve the following linear programming problem.

Min. $Z_D = 2x_1 + 2x_2$

s.t. $2x_1 + 4x_2 \geq 1$

$2x_1 + x_2 \geq 1$

$x_1 + 2x_2 \geq 1$

$x_1, x_2 \geq 0.$

Solution:

The dual of this problem is given by

Max. $Z_D = w_1 + w_2 + w_3$

s.t. $2w_1 + 2w_2 + w_3 \leq 2$

$4w_1 + w_2 + 2w_3 \leq 3$

Introducing the slack variable w_{1s} and w_{2s}. The dual problem can be written as

$$\text{Max. } Z_D = w_1 + w_2 + w_3$$

$$\text{s.t. } 2w_1 + 2w_2 + w_2 + w_{1s} = 2$$

$$4w_1 + w_2 + 2w_3 + w_{2s} = 3$$

$$w_1, w_2, w_{\cdot 1s}, w_{2s} \geq 0$$

Taking $w_1 = 0 = w_2 = w_3$, we have $w_{1s} = 2$ $w_{2s} = 2$. Which is the starting basic feasible solution of the dual.

First Simplex Table

		c_J	1	1	1	0	0	w_B
B	c_B	w_B	Y_1	Y_2	Y_3	Y_{1s}	Y_{2s}	Y_1
x_{1s}	0	2	2	2	1	1	0	1
x_{2s}	0	2	4	1	2	0	1	1/2 mini
	$Z_J - c_J$		–1	–1	–1	0	0	

Proceeding as usual the second simplex table.

Second Table

		c_J	1	1	1	0	0	w_B
B	c_B	w_B	Y_1	Y_2	Y_3	Y_{1s}	Y_{2s}	y
x_{1s}	0	1	0	3/2	0	1	–1/2	x
x_1	1	1/2	1	1/4	1/2	0	1/4	1 mini
			0	–3/4	–1/2	0	1/4	

Proceeding as usual third simplex table.

Third Table

		c_J	1	1	1	0	0	w_B
B	c_B	w_B	Y_1	y_2	Y_3	Y_{1s}	Y_{2s}	y_2
x_{1s}	0	1	0	3/2	0	1	–1/2	2/3 Mini
x_3	1	1	2	1/2	1	0	1/2	2
	$Z_J - c_J$		1	–1/2	0	0	1/2	
				↑		↓		

Proceeding as usual fourth simplex table.

Fourth Table

B	c_B	c_J / w_B	1 Y_1	1 Y_2	1 Y_3	0 Y_{1s}	0 Y_{2s}
x_2	1	2/3	0	1	0	2/3	–1/3
x_3	1	2/3	2	0	1	–1/3	2/3
		$Z_J - c_J$	1	0	0	1/3	1/3

Since all $Z_J - c_J \geq 0$. So solution is optimal.

Solution of primal problem

$x_1 = 1/3$

$x_2 = 1/3$

$Z = 4/3$

For primal constraints, read across the table while for dual constraints read down the columns.

Example 9:

Write the dual and solve the primal problem by simplex method.

$$\text{Max. } Z = 2x_1 - x_2$$

$$\text{s.t.} \quad x_1 + x_2 \leq 10$$

$$4x_1 + 3x_2 \geq 12$$

$$-2x_1 + x_2 \leq 2$$

$$x_1, x_2 \geq 0.$$

Solution:

In standard form the problem can be written as

$$\text{Max. } Z = 2x_1 - x_2$$

$$\text{s.t.} \quad x_1 + x_2 \leq 10$$

$$-(4x_1 + 3x_2) \leq -12$$

$$-2x_1 + x_2 \leq 2$$

$$x_1, x_2 \geq 0$$

The dual of this problem can be written as

$$\text{Mini. } Z_D = 10w_1 - 12w_2 + 2w_2$$

s.t. $w_1 - 4w_2 - 2w_3 \geq 2$

$w_1 - 3w_2 + w_2 \geq -1$

Introducing the slack variable in first and third constraint and introducing surplus and artificial variable is second constraint of the primal problem.

Max. $Z = 2x_1 - x_2$

s.t. $x_1 + x_2 + x_{1s} = 10$

$4x_1 + 3x_2 - x_{2s} + x_{2a} = 12$

$-2x_1 + x_2 + x_{3s} = 2$

Taking $x_1 = 0 = x_2 = x_{2s}$, we have $x_{1s} = 10$, $x_{2a} = 12$ $x_{3s} = 2$ which is the starting basic feasible solution of the primal

First Table

		c_J	2	–1	0	0	0	–M	x_B
B	c_B	x_B	Y_1	Y\	Y_{1s}	Y_{2s}	Y_{3s}	Y_{2a}	Y_1
Y_{4s}	0	10	1	1	1	0	0	0	10
x_{2a}	–M	12	4	3	0	–1	0	1	3mini
x_{3s}	0	2	–2	1	0	0	1	0	–ve
		$Z_J - c_J$	–4M–2 ↑	--3m+1	0	M	0	0 ↓	

Since $\Delta_1 = -4_m - 2$ is mini so y_1 is incoming vector and mini ratio rule out going vector is y_{2a}.

The key element is $y_{21} = 4$

Proceeding as usual the second simplex table.

Second Table

		c_J	2	–1	0	0	0	–M	y_B
B	c_B	x_B	Y_1	Y_2	Y_{1s}	Y_{2s}	Y_{3s}	Y_{2a}	Y_{2s}
x_{1s}	0	7	0	1/4	1	1/4	0	–1/4	21
x_1	2	3	1	3/4	0	–1/4	0	1/4	–ve
x_{3s}	0	8	0	5/2	0	–1/2	1	1/2	–ve
		$Z_J - c_J$	0	5/2	0	–1/2 ↑	0	$\frac{1+2M}{2}$	

Since $\Delta_4 = -1/2$ is mini so y_{2s} is incoming vector and mini ratio rule out going vector is y_{1s}.

The key element is $y_{14} = 1/4$

Proceeding as usual the third simplex table

Third Table

	c_J	2	–1	0	0	0	–M	
B	c_B	x_B	Y_1	Y_2	Y_{1s}	Y_{2s}	Y_{3s}	Y_{2a}
x_{2s}	0	28	0	1	4	1	0	–1
x_1	2	10	1	1	1	0	0	0
x_{3s}	0	22	0	3	2	0	1	0
		Z_J–c_J	0	1	2	0	0	0

Since all $Z_J - c_J \geq 0$, so solution is optimal

$$x_1 = 10, x_2 = 0$$

Max. Z = 20.

Example 10:

Find the dual of problem

$$Max.\quad Z = -x_1 + 2x_2 - x_3$$

$$s.t.\quad 3x_1 + x_2 - x_3 \leq \quad 10$$

$$-x_1 + 4x_2 + x_3 \geq 6$$

$$x_2 + x_3 \leq 4$$

$$x_1, x_2, x_3 \geq 0$$

Solve the primal problem by simplex method and deduce the solution of the problem from the optimal tableaus of the primal.

Solution:

In standard form the problem can be written as

$$\text{Max. } Z_P = -x_1 + 2x_2 - x_3$$

$$3x_1 + x_2 - x_3 \leq 10$$

$$x_2 - 4x_2 - x_3 \leq 6$$

$$x_2 + x_3 \leq 4$$

$$x_1, x_2, x_3, x_4 \geq 0$$

The dual of this problem can be written as

$$\text{Mini. } Z_D = 10w_1 + 6w_2 + 4w_3$$

$$3w_1 \geq -1$$

$$w_1 + w_2 + w_3 \geq 2$$
$$- w_1 - w_2 + w_3 \geq - 1$$

Introducing the slack variable in first and third constraint and surplus and artificial is second constraint of the primal problem. The problem can be written as

$$3x_1 + x_2 - x_3 + x_{1s} = 10$$
$$- x_1 + 4x_2 + x_3 - x_{2s} + x_{2a} = 6$$
$$x_2 + x_3 + x_{3s} = 4$$
$$x_1, x_2, x_3, \ldots x_{3s} \geq 0$$

Taking $x_1 = 0 = x_2 = x_3 = x_{2s}$, we have $x_{1s} = 10$, $x_{2a} = 6$, $x_{3s} = 4$, which is the starting basic feasible solution of the primal.

First Table

		c_J	-1	2	-1	0	0	0	$-M$	x_B
B	c_B	x_B	Y_1	Y_2	Y_3	Y_{1s}	Y_{2s}	Y_{3s}	Y_{2a}	Y_2
x_{1s}	0	10	3	1	-1	1	0	0	0	10
x_{2a}	$-M$	6	-1	4	1	0	-1	0	1	3/2 mini→
x_{3s}	0	4	0	1	1	0	0	1	0	4
		$Z_J - c_J$	M+1	$-4+2$ ↑	$-M+1$	0	M	0	0 ↓	

Since $\Delta_2 = - 4M - 2$ is mini so y_2 is incoming vector and by mini ratio rule out going vector is y_{2a}.

The key element is $y_{22} = 4$

Proceeding as before the second simplex table

Second Table

		c_J	-1	2	1	0	0	0	$-M$	x_B
B	c_B	x_B	Y_1	Y_2	Y_3	Y_{1s}	Y_{2s}	Y_{3s}	Y_{2a}	Y_{2s}
x_{1s}	0	17/2	13/4	0	$-5/4$	1	1/4	$-1/4$	0	34
x_2	2	3/2	$-1/4$	1	1/4	0	$-1/4$	1/4	0	$-$ve
x_{3s}	0	5/2	1/4	0	3/4	0	1/4	$-1/4$	1	10 mini
		$Z_J - c_J$	1/2	0	11/8	0	$-1/2$	$\frac{1}{2} + M$	0	

Proceeding as before the third simplex table

Third Table

		c_J	−1	2	−1	0	0	0	−M
B	c_B	x_B	Y_1	Y_2	Y_3	Y_{1s}	Y_{2s}	Y_{3s}	Y_{2a}
x_{1s}	0	29/4	3	0	−2	1	0	−1	0
x_2	2	4	0	1	1	0	0	1	0
x_{2s}	0	10	1	0	3	0	1	4	−1
		$Z_J - c_J$1	0	3	0	0	2	M	

Since all $Z_J - c_J \geq 0$, so this solution is optimal.

Optimal solution is

$$x_1 = 0 = x_3, \ x_2 = 4$$

Max. Z = 8

solution of the dual

$$w_1 = 0 = w_2, \ w_3 = 2$$

Mini. Z = 8.

Example 11:

Consider the problem,

$$Max.\ Z = 8x_1 + 6x_2$$

$$s.t. \quad x_1 - x_2 \leq 3/5$$

$$x_1 - x_2 \geq 2$$

$$x_1, x_2 \geq 0$$

Show that the primal and dual problem have no feasible solution.

Solution:

In standard form the problem can be written as

$$\text{Max. } Z = 8x_1 + 6x_2$$

$$\text{s.t.} \quad x_1 - x_2 \leq 3/5$$

$$-x_1 + x_2 \leq -2$$

$$x_1, x_2 \geq 0$$

The dual of the this problem can be written as

$$\text{Mini. } Z_D = 3/5\ w_1 - w_2$$

$w_1 - w_2 \geq 8$

$-w_1 + w_2 \geq 6$

$w_1, w_2 \geq 0$

Introducing the slack variable in fist constraint and surplus and artificial variable in second constraint of the primal.

The problem can be written as

Max. $Z = 8x_1 + 6x_2$

s.t. $x_1 - x_2 + x_{1s} = 3/5$

$x_1 - x_2 - x_{1s} + x_{2a} = 2$

$x_1, \ldots, x_{2a} \geq 0$

Taking $x_1 = x_2 = x_{2s}$ we have $x_{1s} = 3/5$ $x_{2a} = 2$. Which is starting basic feasible solution of the primal.

First Table

		c_J	8	6	0	0	–M
B	c_B	x_B	Y_1	Y_2	Y_{1s}	Y_{2s}	Y_{2a}
x_{1s}	0	3/5	1	–1	1	0	0
x_{2a}	–M	2	1	–1	0	–1	1
		$Z_J–c_J$	–M–8	+M–6	0	M	0

Proceeding as before the second simplex table.

Second Table

		c_J	8	6	0	0	–M	x_B
B	c_B	x_B	Y_1	Y_2	Y_{1s}	Y_{2s}	Y_{2a}	Y_2
c_1	8	3/5	1	–1	1	0	0	–ve
x_{2a}	–M	7/5	0	0	–1	–1	1	∞
	$Z_J–c_J$	0	–2 ↑	M+8	M	0		

Every value of $\frac{x_B}{y_2}$ are –ve and ∞, so the primal has no feasible solution and dual has also no feasible solution.

Example 12:

Write the dual of this problem

Max. $Z = 2x_1 + 3x_2$

s.t. $2x_1 + 2x_2 \leq 10$

$x_1 + 2x_2 \leq 6$

$2x_1 + x_2 \leq 6$

$x_1, x_2 \geq 0.$

Solve the primal by the simplex method.

Solution:

The due of this problem can be written as

Mini. $Z_D = 10w_1 + 6w_2 + 6w_3$

$2w_1 + w_2 + 2w_3 \geq 2$

$2w_1 + 2w_2 + w_3 \geq 3$

$w_1, \ldots, w_3 \geq 0$

Introducing the slack variable x_{1s}, x_{2s}, x_{3s} in the primal problem. The problem can be written as

Max. $Z = 2x_1 + 3x_2$

s.t. $2x_1 + 2x_2 + x_{1s} = 10$

$x_1 + 2x_2 + x_{2s} = 6$

$2x_1 + x_2 + x_{3s} = 6$

$x_1, \ldots, x_{3s} \geq 0$

Taking $x_1 = 0 = x_2$ have $x_{1s} = 10, x_{2s} = 6, x_{3s} = 6$

Which is the starting basic feasible solution of the primal.

First Table

		c_J	2	2	0	0	0	x_B
B	c_B	x_B	Y_1	Y_2	Y_{1s}	Y_{2s}	Y_{3s}	Y_1
x_{1s}	0	10	2	2	1	0	0	5
x_{2s}	0	6	1	2	0	1	0	6
x_{3s}	0	6	2	1	0	0	1	2 mini
		$Z_J - c_J$	–2	–2	0	0	0	
			↑				↓	

Proceeding as before the second simplex table.

Second Table

B	c_B	c_J / x_B	2 / Y_1	2 / Y_2	0 / Y_{1s}	0 / Y_{2s}	0 / Y_{3s}	$\frac{x_B}{Y_2}$
x_{1s}	0	4	0	1	1	0	–1	4
x_{2s}	0	3	0	3/2	0	1	–1/2	6
x_1	2	3	1	1/2	0	0	1/2	6
			0	–1	0	0	1	
				↑		↓		

Proceeding as before the third simplex table.

Third Table

B	c_B	c_J / x_B	Y_1	Y_2	Y_{1s}	Y_{2s}	Y_{3s}
x_{1s}	0	2	0	2	1	2/3	–2/3
x_2	3	2	0	1	0	2/3	–1/3
x_1	2	2	1	0	0	–1/3	2/3
	$Z_J–c_J$		0	1	0	4/3	1/3

Since all $Z_J - c_J \geq 0$, so this solution is optimal. The solution of the primal problem is

$x_1 = 2 = x_2$

Max $Z_P = 10$

The solution of dual $w_1 = 0$

$w_2 = 4/3$

$w_3 = 1/3$

Mini. $Z_D = 10$

Example 13:

Use Dual problem to solve the L.P.P.

Max. $Z = 5x_1 - 2x_2 + 3x_3$

s.t. $2x_1 + 2x_2 - x_3 \geq 2$

$3x_1 - 4x_2 \leq 3$

$x_2 + 2x_3 \geq 5$

$x_1, x_2, x_3 \geq 0.$

Solution:

In the standard form the problem is written as

$$\text{Max. } Z = 5x_1 - 2x_2 + 3x_3$$

$$\text{s.t.} \quad 2x_1 - 2x_2 + x_3 \leq -2$$

$$3x_1 - 4x_2 \leq 3$$

$$x_2 + 3x_3 \leq 5$$

$$x_1, x_2, x_3 \leq 0$$

The dual of this problem is given by

$$\text{Min } Z_D = -2w_1 + 3w_2 + 5w_3$$

$$\text{s.t.} \quad -2w_1 + 3w_2 \geq 5$$

$$-2w_1 - 4w_2 + w_3 \geq -2$$

$$w_1 + 3w_3 \geq 3$$

$$w_1, w_2, w_3 \geq 0$$

Multiply the second constraint by – 1

$$2w_2 + 4w_2 - w_3 \leq 2$$

Introducing the surplus and artificial variable in first and third constraint and slack variable in second constraint. The dual problem reduce to the following form.

$$\text{Max. } Z_D = 2w_1 - 3w_2 - 5w_3$$

$$\text{s.t.} \quad -2w_1 + 3w_2 - w_{1s} + w_{1a} = 5$$

$$2w_1 + w_2 - w_3 + w_{2s} = 2$$

$$w_1 + 3w_3 - w_{3s} + w_{3a} = 3$$

$$w_1, w_2, \ldots, w_{3a} \geq 0$$

Taking $w_1 = 0 = w_2 = w_3 = w_{1s} = w_{3s}$, we have $w_{1a} = 5$, $w_{2s} = 2$, $w_{3a} = 3$

Which is the starting basic feasible solution of the dual.

First Table

		c_J	2	–3	–5	0	0	0	–M	–M	w_B
B	c_B	w_B	Y_1	Y_2	Y_3	Y_{1s}	Y_{2s}	Y_{3s}	Y_{1a}	Y_{3a}	Y_2
x_{1a}	–M	5	–2	3	0	–1	0	0	1	0	5/3
x_{2s}	0	2	2	4	–1	0	1	0	0	0	2/4mini→
x_{3a}	–M	3	1	0	3	0	0	–1	0	1	∞
		Z_J–c_J	M–2	–3M+3 ↑	–3M+5	M	0 ↓	M	0	0	

Proceeding as before second simplex table

Second Table

		c_J	2	–3	–5	0	0	0	–M	–M	w_B
B	c_B	w_B	Y_1	Y_2	Y_3	Y_{1s}	Y_{2s}	Y_{3s}	Y_{1a}	Y_{2a}	Y_3
x_{1a}	–M	7/2	–7/2	0	3/4	–1	–3/4	0	1	0	14/3
x_2	–3	1/2	1/	1	–1/4	0	1/4	0	0	0	–ve
x_{3s}	–M	3	1	0	3	0	0	–1	0	1	1 mini
		$Z_J–c_J$	$\frac{5M-7}{2}$	0	$\frac{23-15M}{2_1}$ ↑	M	$\frac{3M-3}{4}$	M	0	0 ↓	

Proceeding as usual third simplex table

Third Table

		c_J	2	–3	–5	0	0	0	–M	–M	w_B
B	c_B	w_B	Y_1	Y_2	Y_3	Y_{1s}	Y_{2s}	Y_{3s}	Y_{1a}	Y_{3a}	
x_{1a}	–M	11/4	–15/4	0	0	–1	–3/4	(1/4)	1	–1/4	11 mini
x_2	–3	3/4	7/12	1	0	0	1/4	–1/12	0	1/12	–ve
x_3	–5	1	1/3	0	1	0	0	–1/3	0	1/3	–ve
			$\frac{45M-95}{12}$	0	0	–M	$\frac{3M-3}{4}$ ↑	0	$\frac{5M-6}{4}$ ↓		

Proceeding as usual fourth simplex table

Fourth Table

		c_J	2	–3	–5	0	0	0
B	c_B	w_B	y_1	Y_2	Y_3	Y_{1s}	Y_{2s}	Y_{3s}
x_{3s}	0	11	–15	0	0	–4	–3	1
x_2	–3	5/3	–2/3	1	0	–1/3	0	0
x_3	–5	14/3	–14/3	0	1	–4/3	–1	0
		$Z_J – c_J$0	0	0	23/3	5	0	

Since all $Z_J – c_J \geq 0$, so the solution is optimal.

∴ Solution of the primal problem is

$x_1 = 23/3$ $x_3 = 0$

$x_2 = 5$

$$\text{Max. } Z_P = \frac{85}{3}.$$

Example 14:

Using the dual, solve the L.P.P.

Mini. $Z_P = 3x_1 + x_2$

s.t. $2x_1 + 3x_2 \geq 2$

$x_1 + x_2 \geq 1$

$x_1, x_2, \geq 0.$

Solution:

The dual of the given L.P.P. is

Max. $Z_D = 2w_1 + w_2$

s.t. $2w_1 + w_2 \leq 3$

$3w_1 + w_2 \leq 1$

$w_1, w_2, \geq 0$

Introducing the slack variable w_{1s} and w_{2s} respectively. The dual problem can be written as

Max. $Z_D = 2w_1 + w_2$

s.t. $2w_1 + w_2 + w_{1s} = 3$

$3_{w1} = w_2 + w_{2s} = 1$

Taking $w_1 = 0 = w_2$, we have $w_{1s} = 3$ and $w_{2s} = 1$. Which is the starting basic feasible solution of the dual.

First Simplex Table

		c_J	2	1	0	0	w_B
B	c_B	w_B	Y_1	Y_2	Y_{1s}	Y_{2a}	Y_{i1}
x_{1s}	0	3	2	1	1	0	3/2
x_{2s}	0	1	3	1	0	1	1/3→mini
	Z_J	$- c_J$	–2	–1	0	0	
	↑					↓	
	incoming vector					outgoing vector	

$\Delta_1 = Z_1 - c_1 = c_B\ Y_1 - c_1 = [(0 \times 2 = 0 \times 3) - 2] = -2$
$\Delta_2 = Z_2 - c_2 = c_B\ y_2 - c_2 = [0 \times 1 + 0 \times 1] - 1 = -1$
$\Delta_3 = Z_3 - c_3 = c_B\ y_3 - c_3 = 0$
$\Delta_4 = 0$

To find incoming vector

Since $\Delta_1 = -1$ is mini. of $\Delta_1, \Delta_2, \Delta_3, \Delta_4$. So y_1 is incoming vector 0.

To find outgoing vector

Since y_1 is incoming vector, so we find ratio

$$= \text{Mini}\left\{\frac{w_B}{y_{i1}}\right\}$$

$$= \text{Mini.}\left\{\frac{w_B}{y_{11}}, \frac{w_B}{y_{21}}\right\}$$

$$= \text{Mini} \qquad \left\{\frac{3}{2}, \frac{1}{3}\right\}$$

$\frac{1}{3}$ is mini so x_{2s} is out going vector.

The key element is $y_{21} = 3$.

Firstly, we divide the second marked row of the first table by 3 to get the second row of the second table.

Now, Multiply the new second row by 2 (number in first row and marked column of the first table).

This will give

$$\frac{1}{3}\times 2 \qquad 1 \times 2 \qquad \frac{1}{3}\times 2 \qquad 0 \times 2 \qquad \frac{1}{3}\times 2$$

$$\frac{2}{3} \qquad 2 \qquad \frac{2}{3} \qquad 0 \qquad \frac{2}{3}.$$

Now subtracting these from the corresponding element of the first row of the first table. This will give

$$3-\frac{2}{3} \qquad 2-2 \qquad 1-\frac{2}{3} \qquad 1-0 \qquad 0-\frac{2}{3}$$

$$\frac{7}{3} \qquad 0 \qquad \frac{1}{3} \qquad 1 \qquad -\frac{2}{3}.$$

This is the first row of the second table.

Second Table

B	c_B	c_J / w_B	2 / Y_1	1 / Y_2	0 / Y_{1s}	0 / Y_{2s}	$\frac{w_B}{Y_{12}}$
x_{1s}	0	7/3	0	1/3	1	–2/3	7
x_1	2	1/3	1	1/3	0	1/3	1 mini→
			0	–1/3	0	2/3	
			↓ outgoing vector		↑ incoming vector		

$\Delta_1 = Z_1 - c_1 = 0$

$\Delta_2 = Z_2 - c_2 = -1/3$

$\Delta_3 = Z_3 - c_3 = 0$

$\Delta_4 = Z_4 - c_4 = 2/3$

To find incoming vector

Since $\Delta_2 = -\frac{1}{3}$ is mini of all. so y_2 is incoming vector

To find out vector.

Since y_2 is incoming vector. So we will find ratio Mini. $\left\{\frac{w_B}{y_{i2}}\right\}$

$$= \text{Mini}\left\{\frac{w_B}{y_{12}}, \frac{w}{y_{22}}\right\}$$

$$= \text{Mini}\left\{\frac{7 \times 3}{3 \times 1}, \frac{1 \times 3}{3 \times 1}\right\}$$

$$= \text{Mini}\ \{7, 1\}$$

1 is mini so x_1 is out going vector.

The key element is $y_{22} = 1/3$.

First we divide the second marked row of the second table by 1/3 to get the new second row of the third table.

Now multiply the new second row by 1/3 (number in the first row and marked column of the second table.

This will give

$1 \times \frac{1}{3}$ $\quad 3 \times \frac{1}{3}$ $\quad 1 \times \frac{1}{3}$ $\quad 0 \times \frac{1}{3}$ $\quad 1 \times \frac{1}{3}$

$\frac{1}{3}$ $\quad 1$ $\quad \frac{1}{3}$ $\quad 0$ $\quad \frac{1}{3}$

Now subtracting these from the corresponding element of the first row of the second table. This will give

$\frac{7}{3}-\frac{1}{3}$ $\quad 0 - 1$ $\quad \frac{1}{3}-\frac{1}{3}$ $\quad 1 - 0$ $\quad \frac{-2}{3} \frac{-1}{3}$

2 $\quad -1$ $\quad 0$ $\quad 1$ $\quad -1.$

This is the first row of the third table.

Third Table

		c_J	2	1	0	0
B	c_B	w_B	Y_1	Y_2	Y_{1s}	Y_{2s}
x_{1s}	0	2	–1	0	1	–1
x_2	1	1	3	1	0	1
		$Z_J - c_J$	1	0	0	1

Since all $Z_J - c_J \geq 0$, so this solution is optimal.

Optimal solution of the dual problem is

$$w_1 = 0$$
$$w_2 = 1$$
$$\text{Max } Z_D = 1$$

solution of original problem

Form the find table solution of the primal is given by

$$x_1 = 0$$
$$x_2 = 1$$
$$\text{Mini. } Z_P = \text{Max } Z_D = 1$$

Example 15:

Using the dual solve the L.P.P.

Mini $\quad Z_P = 10x_1 + 6x_2 + 2x_3$

s.t. $\quad -x_1 + x_2 + x_3 \geq 1$

$3x_1 + x_2 - x_3 \geq 2$

$x_1, x_2, x_3, x_4 \geq 0.$

Solution:

The dual of this problem is given by

Max $Z_D = w_1 + 2w_2$

s.t. $-w_1 + 3w_2 \leq 10$

$w_1 + w_2 \leq 6$

$w_1 - w_2 \leq 2$

$w_1, w_2 \geq 0.$

Introducing the slack variable w_{1s}, w_{2s} and w_{3s} respectively. The dual problem can be written as

Max $Z_D = w_1 + 2w_2$

s.t. $-w_1 + 3w_2 + w_{1s} = 10$

$w_1 + w_2 + w_{2s} = 6$

$w_1 - w_2 + w_{3s} = 2$

Taking $w_1 = 0 = w_2$, we have $w_{1s} = 10$, $w_{2s} = 6$ and $w_{3s} = 2$. Which is the starting basic feasible solution of the dual

First Simplex Table

		c_J	1	2	0	0	0	w_B
B	c_B	w_B	Y_1	Y_2	Y_{1s}	Y_{2s}	Y_{3s}	Y_{12}
x_{1s}	0	10	–1	3	1	0	0	10/3mini→
x_{2s}	0	6	1	1	0	1	0	6/1
x_{3s}	0	2	1	–1	0	0	1	–ve
			–1	–2	0	0	0	
				↑	↓			

Calculate of $Z_J - c_J$

$\Delta_1 = Z_1 - c_1 = -1,$

$\Delta_2 = -2$

To find incoming vector

Since $\Delta_2 = -2$ is mini of Δ_1 and Δ_2.

So y_2 is incoming vector.

To find out going vector.

Since y_2 is incoming vector so we will find ratio Mini $\left\{\frac{w_B}{y_{i2}}, y_{i2} > 0\right\}$

$$= \text{Mini}\left\{\frac{w_{B1}}{y_{12}}, \frac{w_{B2}}{y_{22}}, \frac{w_{B3}}{y_{32}}\right\}$$

$$= \text{Mini}\left\{\frac{10}{3}, \frac{6}{1}, \frac{2}{-1}\right\}$$

$\frac{10}{3}$ is mini so x_{1s} in out going.

The key element is $y_{12} = 3$

Firstly, we divide the first marked row of the first table by 3 to get new first row of the second table. Calculating second and third row from first row.

Second Table

		c_J	1	2	0	0	0	w_B
B	c_B	w_B	Y_1	Y_2	Y_{1s}	Y_{2s}	Y_{3s}	y_{11}
x_2	2	10/3	–1/3	1	1/3	0	0	–ve
x_{2s}	0	8/3	4/3	0	–1/3	1	0	2Mini→
x_{3s}	0	16/3	2/3	0	1/3	0	1	8
			–5/3	0	2/3	0	0	
			↑			↓		

calculate of $Z_J - c_J$

$\Delta_1 = Z_1 - c_1 = -5/3,$

$\Delta_2 = Z_2 - c_2 = 0$

$\Delta_2 = 2_3 - c_3 = 2/3,$

To find incoming vector

Since $\Delta_1 = -5/3$ is mini of Δ_1, Δ_2 Δ_3. So y_1 is incoming vector

To find out going vector.

Since y_1 is incoming vector so we will find ratio Mini $\left\{\frac{w_B}{y_{11}}, y_{11} > 0\right\}$

$$= \text{Mini}\left\{\frac{w_{B1}}{y_{11}}, \frac{w_{B2}}{y_{21}}, \frac{w_{B3}}{y_{32}}\right\}$$

$$= \text{Mini} \left\{ -\text{ve}, \frac{8\times 3}{3\times 4}, \frac{16\times 3}{3\times 2} \right\}$$

2 is mini so x_{2s} is out going vector.

The key element is $y_{21} = 4/3$

First we divide the second marked row of the second table by 4/3 to get the new second of the third table.

Calculating first and third row from second row.

Third Simplex Table

		c_J	1	2	0	0	0
B	c_B	w_B	Y_1	Y_2	Y_{1s}	Y_{2s}	Y_{3s}
x_2	2	4	0	1	1/4	1/4	0
x_1	1	2	1	0	–1/4	3/4	0
x_{3s}	0	2_1	0	0	1/2	–1/2	1
			0	0	1/4	5/4	0

All $Z_J - c_J \geq 0$, so this solution is optimal.

Optimal solution of the dual problem is

$w_1 = 2$

$w_2 = 4$

Max $Z_D = 10$

Solution of the original problem is

$x_1 = 1/4$

$x_2 = 5/4$

$x_3 = 0$

Mini $Z_P = 10.$

Example 16:

Using the dual, solve the following L.P.P.

Max $Z_P = 3x_1 - 2x_2$

s.t. $x_1 \leq 4$

$x_2 \leq 6$

$x_1 + x_2 \leq 5$

$-x_2 \leq -1$

$x_1, x_2 \geq 0.$

Solution:

The dual of the given problem is given by

Mini $Z_D = 4w_1 + 6w_2 + 5w_3 - w_4$

s.t. $w_1 + w_3 \geq 3$

$w_2 + w_3 - w_4 \geq -2$

and $w_1, w_2, w_3, w_4 \geq 0.$

Now firstly we cell convert the dual problem of minimization to maximization problem by taking the objective function as

Max $Z_D' = -Z_D = -4w_1 - 6w_2 - 5w_3 + w_4$

s.t. $w_1 + w_3 \geq 3$

$-w_2 - w_3 + w_4 \leq 2$

Introducing the surplus, slack variable w_{1s}, w_{2s}, respectively, the constraints of the dual problem reduces to the following equities.

$w_1 + w_3 - w_{1s} = 3$

$-w_2 - w_3 + w_4 + w_{2s} = 2$

$w_1, w_2, w_3, w_4, w_{1s}, w_{2s} \geq 0.$

Since identity matrix I_2 is present in the coefficient matrix. So there is no need to introduce artificial variable in the first constraint.

Taking $w_2 = 0 = w_3 = w_4 = w_{1s}$, we have $w_1 = 3$ and $w_{2s} = 2$. Which is the starting basic feasible solution of the dual.

First Table

		c_J	−4	−6	−5	1	0	0	$\frac{w_B}{Y_{14}}$
B	c_B	w_B	Y_1	Y_2	Y_3	Y_4	Y_{1s}	Y_{2s}	
x_1	−4	3	1	0	1	0	−1	0	$\frac{3}{0} = \infty$
x_{2s}	0	2	0	−1	−1	1	0	1	2 mini→
	$Z_J - c_J$		0	6	1	−1	4	0	
						↑		↓	

Calculate $Z_J - c_J$

$\Delta_1 = Z_1 - c_1 = 0$, $\Delta_2 = 6$, $\Delta_3 = 1$

$\Delta_4 = -1$, $\Delta_5 = 4$.

Since $\Delta_4 = -1$ is mini of Δ_1 Δ_2 Δ_3 so y_4 is incoming vector.

To find the out going vector since y_4 is incoming vector so we will find ratio Mini $\left\{\frac{w_B}{y_{i4}}, y_{i4} >\right\}$

Mini $\{\infty, 2\}$

2 is mini x_{2s} is out going vector.

The key element is $y_{24} = 1$

Proceeding as before the second simplex table is

Second Simplex Table

		c_J	–4	–6	–5	1	0	0
c_B	w_B	Y_1	Y_2	Y_3	Y_4	Y_{1s}	Y_{2s}	
x_1	–4	3	1	0	1	0	–1	0
x_4	1	2	0	–1	–1	1	0	1
	$Z_J - c_J$		0	5	0	0	4	1

Since all $Z_J - c_J \geq 0$, so this solution of the dual problem is optimal

Optimal solution of the dual problem is

$w_1 = 3$, $w_2 = 0$

$w_3 = 0$, $w_4 = 2$

Mini $\quad Z_D = \text{Max } Z_D' = 10$

Solution of the original problem

$x_1 = 4$

$x_2 = 1$

Max. $\quad Z_P = 10$.

Example 17:

Find the dual of the following problem and hence solve it.

Mini $\quad Z_P = 6x_1 + 5x_2 + 2x_3$

s.t. $\quad x_1 + 3x_2 = 2x_3 \geq 5$

$4x_1 - 2x_2 + 3x_3 \geq -1$

$$2x_1 + 2x_2 + x_3 \geq 2$$

$$x_1, x_2, x_3 \geq 0.$$

Solution:

The dual of this problem is given by

Max. $Z_D = 5w_1 - w_2 + 2w_2$

s.t. $w_1 + 4w_2 + 2w_3 \leq 6$

$3w_1 - 2w_2 + 3w_3 \leq 5$

$2w_1 + 3w_2 + w_3 \leq 2$

$w_1, w_2, w_3 \leq 0.$

Introducing the slack variable w_{1s}, w_{2s}.

Example 18:

Write the dual of following problem and solve it.

Max. $Z = 4x_1 + 2x_2$

subject to $-x_1 - x_2 \leq -3$

$-x_1 + x_2 \leq -2,\ x_1, x_2 \geq 0.$

Hence or otherwise write down the solution of the primal.

Solution:

The given problem is in standard primal form.

$\therefore$ the dual to the given primal is

Min. $Z_D = -3w_1 - 2w_2$

subject to $-w_1 - w_2 \geq 4$

$-w_1 + w_2 \geq 2$

$w_1, w_2 \geq 0.$

Changing the objective function to maximization and introducing surplus variables $w_3 \geq 0$, $w_4 \geq 0$ and artificial variables $w_5, w_6 \geq 0$ the above problem reduces to

Max. $Z_D = 3w_1 + 2w_2 + 0w_3 + 0w_4 - Mw_5 - Mw_6$

subject to $-w_1 - w_2 - w_3 + w_5 = 4$

$-w_1 + w_2 - w_4 + w_6 = 2$

The starting B.F.S. is $w_1 = 0 = w_2 = w_3 = w_3$, $w_5 = 4$, $w_6 = 2$.

Now applying the simplex method to obtain the optimal solution, we have

B	c_B	c_j / w_B	3 / W_1	2 / W_2	0 / W_3	0 / W_4	–M / W_5	–M / W_6	Min. ratio / w_B/W_2
W_5	–M	4	–1	–1	–1	0	1	0	–ive
W_6	–M	2	–1	1	0	–1	0	1	2
$Z_D = -6_M$		Δ_j	3 – 2M	2	–M	–M	0	0	
W_5	–M	6	–2	0	–1	–1	1	1	
W_2	2	2	–1	1	0	–1	0	1	
$Z_D = 4 - 6M$		Δ_j	5 – 2M	0	–M	2 – M	0	–2	

In the last simplex table no $\Delta_j > 0$ and a non-zero artificial variable appears in the basis therefore the dual problem does not possess any optimum basic feasible solution. Consequently, the given problem does not possess any finite optimal solution.

And w_{3s}. The dual problem can be written as

$$\text{Max. } Z_D = 5w_1 - w_2 + 2w_2$$

$$\text{s.t.} \quad w_1 + 4w_2 + 2w_3 + w_{1s} = 6$$

$$3w_1 - 2w_2 + 3w_3 + w_{2s} = 5$$

$$2w_1 + 3w_2 + w_3 + w_{3s} = 2$$

$$w_1, w_2 - w_{3s} \geq 0.$$

Taking $w_1 = 0$ $w_2 = w_3$, we have $w_{1s} = 6$, $w_{2s} = 5$, $w_{3s} = 2$ which is the starting basic feasible solution of the dual.

First Simplex Table

B	c_B	c_J / w_B	5 / Y_1	–1 / Y_2	2 / Y_3	0 / Y_{1s}	0 / Y_{2s}	0 / Y_{3s}	w_B / Y_{11}
x_{1s}	0	6	1	4	2	1	0	0	6/1
x_{2s}	0	5	3	–2	2	0	1	0	5/3
x_{3s}	0	2	2	3	1	0	0	1	1 mini
			–5	1	–2	0	0	0	
			↑					↓	

Calculate of $Z_J - c_J$

$\Delta_1 = -5$, $\Delta_2 = 1$,

$\Delta_3 = -2$

Since $\Delta_1 = -5$ is mini of $\Delta_1, \Delta_2, \Delta_3$ so y_1 is incoming vector

To find out going vector

Since y_1 is incoming vector so we will find ratio Mini. $\left\{\frac{w_B}{y_{i1}}, y_{i1} > 0\right\}$

$$\text{Mini.} \left\{\frac{6}{1}, \frac{5}{3}, \frac{2}{2}\right\}$$

1 is mini so y_{3s} is out going vector.

The key element is $y_{31} = 2$

Proceeding as before the second simplex table

Second Table

		c_J	5	–1	2	0	0	0
B	c_B	w_B	Y_1	y_2	Y_3	Y_{1s}	Y_{2s}	Y_{3s}
x_{1s}	0	5	0	5/2	3/2	1	0	–1/2
x_{2s}	0	2	0	13/2	1/2	0	1	–3/2
x_1	5	1	1	3/2	1/2	0	0	1/2
		$Z_J - C_J$	0	17/2	5/2	0	0	5/2

Since all $Z_J - c_J \geq 0$. Therefore solution is optimal.

i.e., optimal solution of the dual problems

$w_1 = 1,$

$w_2 = 0,$

$w_3 = 0$

Max. $Z_D = 5$

solution of the primal problem is

$x_1 = 0, x_2 = 0$

$x_3 = 5/2$

Mini. $Z_P = 5.$

Example 19:

Formulate the following L.P.P., into, its dual problem and solve it

Max $Z = x_1 - x_2$

s.t., $2x_1 + x_2 \geq 2$

$-x_1 - x_2 \geq 1$

and $x_1, x_2 \leq 0.$

Solution:

The given L.P.P. is maximizations so all sign should be less than

$$-2x_1 - x_2 \leq -2$$

$$x_2 + x_2 \leq -1$$

The dual of this problems is given by

Mini $Z_D = -2w_1 - w_2$

s.t., $-2w_1 + w_2 \geq 1$

$-w_1 + w_2 \geq -1$

Now firstly we will convert the dual problem of minimization to maximizations problem by taking the objective function as

$$\text{Max } Z_D' = -Z_D = 2w_1 + w_2$$

s.t., $-2w_1 + w_2 \geq 1$

$w_1 - w_2 \leq 1$

Introducing the surplus, slack and artificial variable w_{1s}, w_{2s}, w_{1a} respectively. The dual problem can be written as

Max $Z_D' = -Z_D = 2w_1 + w_2$

s.t., $-2w_1 + w_2 - w_{1s} + w_{1a} = 1$

$w_2 - w_2 + w_{2s} = 1$

Taking $w_1 = 0 = w_2 = w_{1s}$, we have $w_{1a} = 1$ $w_{2s} = 1$, which is the starting B.F.S.

First Table

		c_J	2	1	0	0	–M	w_B
B	c_B	w_B	Y_1	Y_2	Y_{1s}	Y_{2s}	Y_{1a}	Y_{i2}
x_{1a}	–M	1	–2	1	–1	0	1	1
x_{2s}	0	1	1	–1	0	1	0	–ive
	$Z_J–c_J$		$2_M–2$ ↑	–M–1	M	0	0 ↓	

$\Delta_1 = 2M - 2,$

$\Delta_2 = -M - 1, \ \Delta_3 = M$

Since $\Delta_2 = -M - 1$ is mini of Δ_1, Δ_3. So y_2 is incoming vector

To find out doing vector.

By mini ratio rule we find that y_{1a} is out going vector.

The key element is 1

Proceeding as before second simplex table

Second Table

B	c_B	c_J / w_B	2 / Y_1	1 / Y_2	0 / Y_{1s}	0 / Y_{2s}	–M / Y_{1a}	$\frac{w_B}{Y_{i3}}$
x_2	1	1	–2	1	–1	0	1	–ive
x_{2s}	0	2	–1	0	(1)	1	1	2→
		0	0	–1	0	1 + M		
				↑	↓			

Proceeding as usual the third table as follows:

Third Table

B	c_B	c_J / w_B	2 / Y_1	1 / Y_2	0 / Y_{1s}	0 / Y_{2s}	–M / Y_{1a}	$\frac{w_B}{Y_{i1}}$
x_2	1	3	–3	1	0	1	2	–ive
x_{1s}	0	2	–1	0	1	1	1	–ive
	$Z_J - c_J$		–5	0	0	1	2+M	
			↑					

Because all value of a are (–ve) therefore solution is no feasible solution.

Example 20:

Give the dual of the following problem and solve:

Mini. $Z_P = x_1 + x_2$

$2x_1 + x_2 \geq 4$

$x_1 + 7x_2 \geq 7$

$x_1, x_2 \geq 0.$

Solution:

The dual of the given problem is given by

Max. $Z_D = 4w_1 + 7w_2$

s.t. $2w_1 + w_2 \leq 1$

$w_1 + 7w_2 \leq 1$

$w_1, w_2 \geq 0$

Introducing the slack variable w_{1s} and w_{2s} respectively. The dual problem can be written as

Max. $Z_D = 4w_1 + 7w_2$

s.t. $2w_1 + w_2 + w_{1s} = 1$

$w_1 + 7w_2 + w_{2s} = 1$

$w_1, w_2, w_{1s}, w_{2s} \geq 0.$

Taking $w_1 = 0 = w_2$, we have $w_{1s} = 1$, $w_{2s} = 1$, which is the starting basic feasible solution of the dual.

First Table

		c_J	4	7	0	0	w_B
B	c_B	w_B	Y_1	Y_2	Y_{1s}	Y_{2s}	Y_{i2}
x_{1s}	0	4	2	1	1	0	4
x_{2s}	0	7	1	7	0	1	1 mini
	$Z_J - c_J$		–4	–7	0	0	

$\Delta_1 = -4$, $\Delta_2 = -7$

Since $\Delta_2 = -7$ is mini of Δ_1 and Δ_2 so y_2 is incoming vector and by minimum ratio rule outgoing vector is y_{2s}.

The key element is 7.

Proceeding as before the second simplex table.

Second Table

		c_J	4	7	0	0	w_B
B	c_B	w_B	Y_1	y_2	Y_{1s}	Y_{2s}	Y_{i1}
x_{1s}	0	3	13/7	0	1	–1/7	21/13 mini
x_2	7	1	1/7	1	0	1/7	7
	$Z_J - c_J$		–3	0	0	1	
			↑		↓		

Since $\Delta_1 = -3$ is mini so y_1 is incoming vector and by mini rule method outgoing vector is y_{1s}.

The key element is $y_{11} = 13/7$

Proceeding as before the third simplex table.

Third Table

B	c_B	c_J / w_B	4 / Y_1	7 / Y_2	0 / Y_{1s}	0 / Y_{2s}
x_1	4	21/13	1	0	7/13	–1/13
x_2	7	10/13	0	1	–1/13	1/13
		$Z_J - c_J$	0	0	$\frac{21}{13}$	10/13

Since all $Z_J - c_J \geq 0$, so this solution of the dual problem is optimal solution of the dual is

$w_1 = 21/13,\ w_2 = 10/13$

$\text{Max } Z_D = \frac{31}{13}$

solution of the primal problem is

$x_1 = \frac{21}{13},\ x_2 = 10/13$

$\text{Mini. } Z_P = 31/13.$

Example 21:

Write the dual of L.P.P. and solve it.

$$\text{Mini } Z = 8x_1 + 2x_2 + 4x_3$$

$$\text{s.t. } x_1 - 4x_2 - 2x_3 \geq 2$$

$$x_1 + x_2 - 3x_3 \geq -1$$

$$-3x_1 - x_2 + x_3 \geq 1$$

$$x_1, x_2, x_3 \geq 0.$$

Solution:

The dual of this problem is given by

$$\text{Max. } Z_D = 2w_1 - w_2 + w_3$$

$$\text{s.t. } w_1 + w_2 - 3w_3 \leq 8$$

$$-4w_1 + w_2 - w_3 \leq 2$$

$$-2w_1 - 3w_2 + w_3 \leq +4$$

$$w_1, w_2, w_3 \geq 0$$

Introducing the slack variable w_{1s}, w_{2s} and w_{3s} respectively. The dual problem can be written as

$$\text{Max } Z_D = 2w_1 - w_2 + w_1$$

$$\text{s.t.} \quad w_1 + w_2 - 3w_3 + w_{1s} = 8$$

$$-4w_1 + w_2 - w_3 + w_{2s} = 2$$

$$-2w_1 - 3w_2 + w_3 + w_{3s} = +4$$

$$w_1, w_2, w_3, w_{1s}, w_{2s}, w_{3s} \geq 0$$

Taking $w_1 = 0 = w_2 = w_3$ we have $w_{1s} = 8$,

$$w_{2s} = 2 \; w_{3s} = 4,$$

which is the starting basic feasible solution of the dual.

First Table

		c_j	2	–1	1	0	0	0	w_B
B	c_B	w_B	Y_1	Y_2	Y_3	Y_{1s}	Y_{2s}	Y_{3s}	Y_{i1}
x_{1s}	0	8	1	1	–3	1	0	0	8
x_{2s}	0	2	–4	1	–1	0	1	0	–ive
x_{3s}	0	4	–2	–3	1	0	0	1	–ive
			–2	1	–1	0	0	0	
		↑			↓				

Since $\Delta_1 = -2$ is mini of $\Delta_1, \Delta_2, \Delta_3$. So y_1 is incoming vector and by mini ratio rule out going vector is y_{1s}.

The key element is $y_{11} = 1$

Proceeding as before the second simplex table

Second Table

		c_J	2	–1	1	0	0	0	w_B
B	c_B	w_B	Y_1	Y_2	Y_3	Y_{1s}	Y_{2s}	Y_{3s}	Y_{i3}
x_1	2	8	1	1	–3	1	0	0	–ive
x_{2s}	0	34	0	5	–13	4	1	0	–ive
x_{3s}	0	20	0	–1	–5	2	0	1	–ive
			0	3	–7	2	0	0	
					↑				

All $\frac{w_B}{y_{i3}}$ is negative so solution is no feasible solution or unbounded solution.

Example 22:

Write the dual of L.P.P. and solve it.

$$\text{Mini } Z_P = 3x_1 - 2x_2 + 4x_3$$

$$\text{s.t. } 3x_1 + 5x_2 + 4x_3 \geq 7$$

$$6x_1 + x_2 + 3x_3 \geq 4$$

$$7x_1 - 2x_2 - x_3 \leq 10$$

$$x_1 - 2x_2 + 5x_3 \geq 3$$

$$4x_1 + 7x_2 - 2x_3 \geq 2$$

$$x_1, x_2, x_3 \geq 0.$$

Solution:

In standard form the problem can be written as

$$\text{Mini } Z_P = 3x_1 + 2x_2 + 4x_2$$

$$\text{s.t. } 3x_1 + 5x_2 + 4x_3 \geq 7$$

$$6x_1 + x_2 + 3x_3 \geq 4$$

$$-x_1 + 2x_2 + x_3 \geq -10$$

$$x_1 - 2x_2 + 5x_3 \geq 3$$

$$4x_1 + 7x_2 - 2x_3 \geq 2$$

$$x_1, x_2, x_3 \geq 0.$$

The dual of this problem can be written as

$$\text{Max } Z_D = 7w_1 + 4w_2 - 10w_3 + 3w_4 + 2w_5$$

$$\text{s.t. } 3w_1 + 6w_2 - 7w_3 + w_4 = 4w_5 \leq 3$$

$$5w_1 + w_2 + 2w_3 - 2w_4 + 7w_5 \leq -2$$

$$4w_1 + 3w_2 + w_3 + 5w_4 - 2w_5 \leq 4$$

$$w_1, w_2, w_3, w_4, w_5 \geq 0$$

Multiply the second constraint by

$$-1 - 5w_1 - w_2 - 2w_3 + 2w_4 - 7w_5 \geq 2$$

Introducing the slack variable w_{1s} and w_{3s} in the first and third constraint introducing and surplus variable and artificial variable w_{2s} an w_{29} in the second constraint, and taking the negative price M to artificial variable, the dual problem reduce to the following form.

$$\text{Max} \quad Z_D = 7w_1 + 4w_2 - 10w_3 + 3w_4 + 2w_5$$

$$\text{s.t.} \quad 3w_1 + 6w_2 - 7w_3 + w_4 + 4w_5 + w_{1s} = 3$$

$$- 5w_1 - w_2 - 2w_3 + 2w_4 - 7w_5 - w_{2s} + w_{2a} = 2$$

$$4w_1 + 3w_2 + w_3 + 5w_4 - 2w_5 + w_{3s} = 4$$

$$w_1, w_2, w_3, w_4, w_5, w_{1s}, w_{2s}, w_{3s}, w_4 \geq 0$$

Taking $w_1 = 0 = w_2 = w_3 = w_4 = w_5 = w_{2s}$,

we have $w_{1s} = 3$, $w_{2a} = 2$, $w_{3s} = 4$. Which is the starting basic feasible solution of the dual.

First Simplex Table

		c_J	7	4	–10	3	2	0	0	0	–M	w_B
B	c_B	w_B	Y_1	Y_2	Y_3	Y_4	Y_5	Y_{1s}	Y_{2s}	Y_{3s}	Y_{2a}	Y_4
x_{1s}	0	3	3	6	–7	1	4	1	0	0	0	3/1
x_{29}	–M	2	–5	–1	–2	2	–7	0	–1	0	1	2/2
x_{3s}	0	4	4	3	1	5	–2	0	0	1	0	4/8mini
		$Z_J–c_Js_M$	–7M–4	10+2_M	–3-2M	7M–2	0	M	0	0		
					↑				↓			

Since $\Delta_4 = -3 - 2_M$ is mini of all. So y_4 is incoming vector and by mini ratio rule out going vector is y_{3s}.

The key element is $y_{34} = 5$

Proceeding as usual the second simplex table.

Second Table

		c_J	7	4	–10	3	2	0	0	0	–M	w_B
B	c_B	w_B	Y_1	Y_2	Y_3	Y_4	Y_5	Y_{1s}	Y_{2s}	Y_{3s}	Y_{2a}	Y
x_{1s}	0	11/5	11/5	27/5	–36/5	22/5	0	1	0	–1/5	0	
x_{2a}	–M	2/5	-33/5	–11/5	–12/5	0	–31/5	0	–1	–2/5	1	
x_4	3	4/5	4/5	3/5	1/5	1	–2/5	0	0	1/5	0	
		$Z_J–c_J$	$\frac{33M-2s}{5}$	$\frac{11M-11}{5}$	$\frac{53+12M}{5}$	0	$\frac{31M-16}{5}$	0	M	$\frac{2M+3}{5}$	0	

Since all $Z_j - c_j \geq 0$, so this solution is optimal but the artificial vector, appear in the basis at the positive level which indicates that the dual problem has no feasible solution, so primal also has no solution.

EXERCISES

1. Prove that if the primal has an unbounded solution, the dual has no solution and *vice-versa.*
2. Define the dual of given L.P.P. prove that dual of a dual is primal problem.
3. Write the dual of the following L.P. problem.

(a) Max. $Z = x_1 + 3x_2$

s.t. $3x_1 + 2x_2 \leq 6$

$3x_1 + 2x_2 = 4$

$x_1, x_2 \geq 0.$

(b) Mini. $Z = 3x_1 + x_2$

s.t. $2x_1 + 3x_2 \geq 2$

$x_1 + x_2 \geq 1$

and $x_1, x_2 \geq 0.$

(c) Max. $Z = 3x_1 - 2x_2$

$x_1 \leq 4$

$x_2 \leq 6$

$x_1 + x_2 \leq 5$

$-x_2 \leq -1$

$x_1, x_2 \geq 0.$

(d) Max. $Z = 3x_1 + 5x_2 + 4x_3$

s.t. $2x_1 + 3x_2 \leq 8$

$2x_2 + 5x_3 \leq 10$

$3x_1 + 2x_2 + 4x_3 \leq 15$

and $x_1, x_2, x_3 \geq 0.$

(e) Mini $Z = 15x_1 + 10x_2$

s.t. $3x_1 + 5x_2 \geq 5$

$5x_1 + 2x_2 \geq 3$

$x_1, x_2 \geq 0$

4. Use the Dual problem to solve the following problem.

(i) Max $Z = 3x_1 + x_2$

s.t. $x_1 + x_2 \leq 1$

$2x_1 + 3x_2 \geq 2$

$x_1, x_2 \geq 0.$

(ii) Mini. $Z_P = x_1 + x_2 + x_3$

$x_1 + 3x_2 + 4x_3 = 5$

$x_1 - 2x_2 \leq 3$

$2x_2 - x_3 \geq 4$

$x_1, x_2 \geq 0$

x_3 is unrestricted.

(iii) Max $Z = 4x_1 + 3x_2$

$x_1 \leq 6$

$x_2 \leq 8$

$x_1 + x_2 \leq 7$

$3x_1 + x_2 \leq 15$

$-x_2 \leq 1$

$x_1, x_2 \geq 0.$

(iv) Mini $Z = 3x_1 - 2x_2 + 4x_3$

s.t. $3x_1 + 5x_2 + 4x_3 \geq 7$

$6x_1 + x_2 + 3x_3 \geq 4$

$7x_1 - 2x_2 - x_3 \leq 10$

$x_1 - 2x_2 + 5x_3 \geq 3$

$4x_1 + 7x_2 - 2x_3 \geq 2$

$x_1, x_2, x_3 \geq 0.$

(v) Mini. $Z = 4x_1 + 3x_2 + 6x_3$

s.t. $x_1 + x_3 \geq 2$

$x_2 + x_3 \geq 5$

$x_1, x_2, x_3 \geq 0.$

(vi) Max. $Z = 2x_1 + 3x_2 + 5x_3$

s.t. $x_1 + x_2 + x_3 \leq 7$

$3x_1 - x_2 + x_3 \leq 5$

$$x_1 + 2x_2 + 2x_3 \le 13$$

$$x_1, x_2, x_3 \ge 0.$$

5. Define the dual of given L.P.P. prove that dual of a dual is primal problem.

UNSYMMETRIC PRIMAL DUAL PROBLEMS

Primal Problem

Find a column vector $\mathbf{x} \in R^n$ which maximizes

$$Z_x = c\,x,\ c \in R^n$$

subject to $A\,x = b,\ b \in R^m$,

$$x \ge 0 \text{ and } A \text{ is an } m \times n \text{ real matrix.}$$

Dual Problem

Find a column vector $w \in R^m$ which minimizes

$$Z_w = b'\,w$$

subject to $A'\,w \ge c'$.

In this case *the dual variables are unrestricted in sign.*

DUAL OF AN L.P.P. WITH MIXED RESTRICTIONS

Sometimes a given L.P.P. contains a mixture of inequalities ($\ge$, $\le$), equations; non-negative variables and unrestricted variables, then to obtain its dual we proceed in the following manner:

(i) If a constraint is an equation (has = sign), replace it by two constraints involving the inequalities going in opposite directions.

For example,

the equation $2x_1 + 5x_2 = 9$ is replaced by

$$2x_1 + 5x_2 \le 9 \qquad \text{...(1)}$$

and $$2x_1 + 5x_2 \ge 9. \qquad \text{...(2)}$$

(ii) If the given problem is of maximization, all constraints should have $\le$ sign. If some constraint has $\ge$ sign, multiply both sides by -1 and make the sign $\le$.

Thus in the above example multiply both sides of (2) by -1 so that it becomes $\quad -2x_1 - 5x_2 \le -9$

Similarly, if the problem is of minimization, all constraints should have $\ge$ sign.

(iii) If there is some unrestricted variable, replace it by the difference of two variables.

Note: The dual variables corresponding to primal equality constraints must by unrestricted in sign and those associated with primal inequalities must be non-negative.

Standard Primal Form

A linear programming problem is said to be in *standard primal form if*

(a) For a maximization problem all the constraints have $\leq$ sign.

(b) For a minimization problem all the constraints have $\geq$ sign.

Example 1:

Write the dual of the following problem:

$$\text{Min. } Z = 2x_2 + 5x_3$$

$$\text{subject } x_1 + x_2 \geq 2$$

$$2x_1 + x_2 + 6x_3 \leq 6$$

$$x_1 - x_2 + 3x_3 = 4$$

$$x_1, x_2, x_3 \geq 0.$$

Solution:

First we shall convert the given problem to standard primal form:

1. Since the problem is of minimization therefore all the constraints should have the sign $\geq$.
2. Multiplying the second constraint by -1, it becomes $- 2x_1 - x_2 - 6x_3 \geq - 6.$
3. Since the third constraint is an equality so replacing it by the following two constrains

$$x_1 - x_2 + 3x_3 \leq 4$$

$$x_1 - x_2 + 3x_3 \geq 4$$

or $$- x_1 + x_2 - 3x_3 \geq - 4$$

and $$x_1 - x_2 + 3x_3 \geq 4.$$

$\therefore$ the given problem in standard primal form is

$$\text{Min. } Z = 0x_1 + 2x_2 + 5x_3$$

subject to $x_1 + x_2 \geq 2$

$$-2x_1 - x_2 - 6x_3 \geq -6$$

$$-x_1 + x_2 - 3x_3 \geq -4$$

$$x_1 - x_2 + 3x_3 \geq 4$$

$$x_1, x_2, x_3 \geq 0.$$

The matrix form of the above problem is

Min. $Z = (0, 2, 5)\,(x_1, x_2, x_3)$

subject to

$$\begin{bmatrix} 1 & 1 & 0 \\ -2 & -1 & -6 \\ -1 & 1 & -3 \\ 1 & -1 & 3 \end{bmatrix} \begin{bmatrix} x_1 \\ x_2 \\ x_3 \end{bmatrix} \geq \begin{bmatrix} 2 \\ -6 \\ -4 \\ 4 \end{bmatrix}$$

Now the dual of the given primal is

Max. $Z_D = (2, -6, -4, 4)\,(y_1, y_2, y_3', y_3'')$

subject to

$$\begin{bmatrix} 1 & -2 & -1 & 1 \\ 1 & -1 & 1 & -1 \\ 0 & -6 & -3 & 3 \end{bmatrix} \begin{bmatrix} y_1 \\ y_2 \\ y_3' \\ y_3'' \end{bmatrix} \leq \begin{bmatrix} 0 \\ 2 \\ 5 \end{bmatrix},$$

$$y_1, y_2, y_3', y_3'' \geq 0$$

or Max. $Z_D = 2y_1 - 6y_2 - 4\,(y_3' - y_3'')$

subject to

$$y_1 - 2y_2 - (y_3' - y_3'') \leq 0$$

$$y_1 - y_2 + y_3' - y_3'' \leq 2$$

$$-6y_2 - 3\,(y_3' - y_3'') \leq 5,$$

$$y_1, y_2, y_3', y_3'' \geq 0.$$

Substituting $y_3 = y_3' - y_3''$, the required dual is

Max. $Z_D = 2y_1 - 6y_2 - 4y_3$

subject to

$$y_1 - 2y_2 - y_3 \leq 0$$

$$y_1 - y_2 + y_3 \leq 2$$

$$-6y_2 - 3y_3 \leq 5,$$

$y_1, y_2 \geq 0$ and y_3 is unrestricted in sign.

Example 2:

Write the dual of the following L.P.P.

$$\text{Max. } Z = 2x_1 + 3x_2 + x_3$$

subject to $4x_1 + 3x_2 + x_3 = 6$

$$x_1 + 2x_2 + 5x_3 = 4,$$

$$x_1, x_2, x_3 \geq 0.$$

Solution:

First we shall convert the given L.P.P. into standard primal form.

Since the given problem is of maximization, so all the constraints should have the sign $\leq$.

The standard primal form of the given L.P.P. is

$$\text{Max. } Z = 2x_1 + 3x_2 + x_3$$

subject to $4x_1 + 3x_2 + x_3 \leq 6$

$$-4x_1 - 3x_2 - x_3 \leq -6$$

$$x_1 + 2x_2 + 5x_3 \leq 4$$

$$-x_1 - 2x_2 - 5x_3 \leq -4,$$

$$x_1, x_2, x_3 \geq 0.$$

The matrix form of the above problem is

$$\text{Max. } Z = (2, 3, 1)\,(x_1, x_2, x_3)$$

subject to

$$\begin{bmatrix} 4 & 3 & 1 \\ -4 & -3 & -1 \\ 1 & 2 & 5 \\ -1 & -2 & -5 \end{bmatrix} \begin{bmatrix} x_1 \\ x_2 \\ x_3 \end{bmatrix} \leq \begin{bmatrix} 6 \\ -6 \\ 4 \\ -4 \end{bmatrix},$$

Now the dual of the given primal is

$$\text{Min. } Z_D = (6, -6, 4, -4)\,(y_1', y_1'', y_2', y_2'')$$

$$= 6\,(y_1' - y_1'') + 4\,(y_2' - y_2'')$$

subject to

$$\begin{bmatrix} 4 & -4 & 1 & -1 \\ 3 & -3 & 2 & -2 \\ 1 & -1 & 5 & -5 \end{bmatrix} \begin{bmatrix} y_1' \\ y_1'' \\ y_2' \\ y_2'' \end{bmatrix} \geq \begin{bmatrix} 2 \\ 3 \\ 1 \end{bmatrix}.$$

or $4\,(y_1' - y_1'') + y_2' - y_2'' \geq 2$

$$3\,(y_1' - y_1'') + 2\,(y_2' - y_2'') \geq 3$$

$(y_1' - y_1'') + 5\,(y_2' - y_2'') \geq 1,$

$y_1', y_1'', y_2', y_2'' \geq 0.$

Hence the required dual is

Max. $Z_D = -5y_1 - 3y_2 + 4y_3$

subject to $-y_1 - y_2 \leq 1$

$3y_1 + 2y_2 + 2y_3 \leq 1$

$-4y_1 - y_3 = 1,$

$y_2, y_3 \geq 0$ y_1 is unrestricted.

Substituting $y_1 = y_1' - y_1''$, $y_2 = y_2 - y_2''$, the required dual is

Min $Z_D = 6y_1 + 4y_2$

subject to $4y_1 + y_2 \geq 2$

$3y_1 + 2y_2 \geq 3$

$y_1 + 5y_2 \geq 1.$

y_1, y_2 are unrestricted in sign.

Example 3:

Write the dual of the following problem

Max. $Z = 3x_1 + 5x_2 + 7x_3$

subject to $x_1 + x_2 + 3x_3 \leq 10$

$4x_1 - x_2 + 2x_3 \geq 15,$

$x_1, x_2 \geq 0$, x_3 *is unrestricted.*

Solution:

First we shall write the given L.P.P. in the standard primal form.

Since the given problem is of maximization therefore all the constraints should have the sign $\leq$.

Substituting $x_3 = x_3' - x_3''$, the standard primal form of the problem is

Max. $Z = 3x_1 + 5x_2 + 7\,(x_3' - x_3'')$

subject to

$x_1 + x_2 + 3x_3' - 3x_3'' \leq 10$

$-4x_1 + x_2 - 2x_3 + 2x_3'' \leq -15$

$x_1, x_2, x_3', x_3'', \geq 0.$

The matrix form of the above problem is

Max. $Z = (3, 5, 7, -7)\,(x_1, x_2, x_3', x_3'')$

subject to

$$\begin{bmatrix} 1 & 1 & 3 & -3 \\ -4 & 1 & -2 & 2 \end{bmatrix} \begin{bmatrix} x_1 \\ x_2 \\ x_3' \\ x_3'' \end{bmatrix} \leq \begin{bmatrix} 10 \\ -15 \end{bmatrix}.$$

Now the dual of the given problem is

Min. $Z_D = (10, -15)\,(y_1, y_2)$

subject to

$$\begin{bmatrix} 1 & -4 \\ 1 & 1 \\ 3 & -2 \\ -3 & 2 \end{bmatrix} \begin{bmatrix} y_1 \\ y_2 \end{bmatrix} \geq \begin{bmatrix} 3 \\ 5 \\ 7 \\ -7 \end{bmatrix}$$

or Min. $Z_D = 10y_1 - 15y_2$

subject to

$y_1 - 4y_2 \geq 3$

$y_1 + y_2 \geq 5$

$3y_1 - 2y_2 \geq 7$

$-3y_1 + 2y_2 \geq -7$ or $3y_1 - 2y_2 \leq 7_1$

$y_1, y_2 \geq 0.$

Thus the required dual is

Min. $Z_D = 10y_1 - 15y_2$

$y_1 - 4y_2 \geq 3$

$y_1 + y_2 \geq 5$

$3y_1 - 2y_2 = 7$

$y_1, y_2 \geq 0.$

Example 4:

Write the dual of the following problem

Min. $Z = 10x_1 + 20x_2$

subject to $3x_1 + 2x_2 \geq 18$

$x_1 + 3x_2 \geq 8$

$2x_1 - x_2 \leq 6$

$x_1, x_2 \geq 0.$

Solution:

First we shall convert the given L.P.P. into standard primal form.

1. Since the given problem is of minimization, therefore all the constraints should have $\geq$ sign.
2. The first two inequalities are in the right direction while the third one is not.
3. Multiplying both sides of the third inequality by -1 it changes to $-2x_1 + x_2 \geq -6$.
4. Thus the standard primal form of the given L.P.P. is

$$\text{Min } Z = 10x_1 + 20x_2$$

$$\text{subject } 3x_1 + 2x_2 \geq 18$$

$$x_1 + x_2 \geq 8$$

$$-2x_1 + x_2 \geq 6.$$

$$\text{and} \quad x_1, x_2 \geq 0.$$

5. The matrix form of the above L.P.P. is

$$\text{Min. } Z = (10, 20)(x_1, x_2)$$

$$= c\,x$$

subject

$$\begin{bmatrix} 3 & 2 \\ 1 & 3 \\ -2 & 1 \end{bmatrix} \begin{bmatrix} x_1 \\ x_2 \end{bmatrix} \geq \begin{bmatrix} 18 \\ 8 \\ -6 \end{bmatrix}$$

or $\quad A\,x \geq b,$

$$x_1, x_2 \geq 0.$$

$\therefore$ the required dual is

$$\text{Max. } Z_D = b'\,y = (18, 8, -6)(y_1, y_2, y_3)$$

$$\text{s.t.} \quad A'\,y = c' \text{ or } \begin{bmatrix} 3 & 1 & -2 \\ 2 & 3 & 1 \end{bmatrix} \begin{bmatrix} y_1 \\ y_2 \\ y_3 \end{bmatrix} \leq \begin{bmatrix} 10 \\ 20 \end{bmatrix}$$

or $\quad$ Max. $Z_D = 18y_1 + 8y_2 - 6y_3$

s.t. $\quad 3y_1 + y_2 - 2y_3 \leq 10$

$2y_1 + 3y_2 + y_3 \leq 20,$

$y_1, y_2, y_3 \geq 0.$

Example 5:

Write the dual of the following problem:

Min. $\quad Z = x_1 + x_2 + x_3$

subject to

$$x_1 - 3x_2 + 4x_3 = 5,$$

$$x_1 - 2x_2 \leq 3,\ 2x_2 - x_3 \geq 4,$$

$x_1, x_2 \geq 0$, x_3 *is unrestricted.*

Solution:

First we shall write the given L.P.P. in the standard primal form, substituting $x_3 = x_3' - x''$, $x_3' \geq 0$, $x_3' \geq 0$. The given problem can be written in the standard primal form as

Min. $Z = x_1 + x_2 + x_3' - x_3$

subject to

$$- x_1 + 3x_2 - 4\,(x_3' - x_3'') \geq -5$$

$$x_1 - 3x_2 + 4\,(x_3' - x_3'') \geq 5$$

$$- x_1 + 2x_2 \geq -3$$

$$2x_2 - (x_3' - x_3'') \geq 4,$$

$$x_1, x_2, x_3', x_3'' \geq 0.$$

The matrix form of the above problem is

Min. $Z = (1, 1, 1, -1)\,(x_1, x_2, x_3', x_3'')$

subject to

$$\begin{bmatrix} -1 & 3 & -4 & 4 \\ 1 & -3 & 4 & -4 \\ -1 & 2 & 0 & 0 \\ 0 & 2 & -1 & 1 \end{bmatrix} \begin{bmatrix} x_1 \\ x_2 \\ x_3' \\ x_3'' \end{bmatrix} \geq \begin{bmatrix} -5 \\ 5 \\ -3 \\ 4 \end{bmatrix}$$

Now the dual of the given primal is

$$\text{Max } Z_D = (-5, 5, -3, 4)\,(y_1', y_1'', y_2, y_3)$$

$$= -5\,(y_1' - y_1'') - 3y_2 + 4y_3$$

subject to
$$\begin{bmatrix} -1 & 1 & -1 & 0 \\ 3 & -3 & 2 & 2 \\ -4 & 4 & 0 & -1 \\ 4 & -4 & 0 & 1 \end{bmatrix} \begin{bmatrix} y_1' \\ y_1'' \\ y_2 \\ y_3 \end{bmatrix} \leq \begin{bmatrix} 1 \\ 1 \\ 1 \\ -1 \end{bmatrix}$$

or
$$-(y_1' - y_1'') - y_2 \leq 1$$
$$3(y_1' - y_1'') + 2y_2 + 2y_3 \leq 1$$
$$-4(y_1' - y_1'') - y_3 \leq 1$$
$$4(y_1' - y_1'') + y_3 \leq -1,$$
$$y_1', y_1'', y_2, y_3 \geq 0.$$

Substituting $y_1 = y_1' - y_1''$, the required dual is

Max. $Z_D = -5y_1 - 3y_2 + 4y_3$

subject to $-y_1 - y_2 \leq 1$

$3y_1 + 2y_2 + 2y_3 \leq 1$

$-4y_1 - y_3 \leq 1$

$4y_1 + y_3 \leq -1$ or $-4y_1 - y_3 \geq 1$,

$y_2, y_3 \geq 0$ and y_1 is unrestricted.

EXERCISES

Find the dual of the following linear programming problems:

1. Min. $Z = 3x_1 + x_2$

   subject to

   $2x_1 + 3x_2 \geq 2$

   $x_1 + x_2 \geq 1,$

   $x_1, x_2 \geq 0.$

2. Min. $Z = 4x_1 + 6x_2 + 18x_3$

   subject to

   $x_1 + 3x_3 \geq 3$

   $x_2 + 2x_3 \geq 5,$

   $x_1, x_2, x_3 \geq 0.$

3. Max. $Z = 3x_1 + 4x_2$

   subject to

$2x_1 + 6x_2 \leq 16$

$5x_1 + 2x_2 \geq 20,$

$x_1, x_2 \geq 0.$

4. Min. $Z = 2x_1 + 3x_2 + 4x_3$

subject to

$2x_1 + 3x_2 + 5x_3 \geq 2$

$3x_1 + x_2 + 7x_3 = 3$

$x_1 + 4x_2 + 6x_3 \leq 5$

$x_1, x_2 \geq 0$, x_3 is unrestricted.

5. Min. $Z = 2x_1 + 2x_2 + 4x_3$

subject to

$2x_1 + 3x_2 = 5x_3 \geq 2$

$3x_1 + x_2 + 7x_3 \leq 3$

$x_1 + 4x_2 + 6x_3 \leq 5,$

$x_1, x_2, x_3 \geq 0.$

6. Min. $Z = x_1 - 3x_2 - 2x_3$

subject to

$3x_1 - x_2 + 2x_3 \leq 7$

$2x_1 - 4x_2 \geq 12$

$-4x_1 + 3x_2 + 8x_3 = 10$

$x_1, x_2 \geq 0$, x_3 is unrestricted.

7. Min. $Z = x_1 - x_2$

subject to

$2x_1 + x_2 \geq 2,$

$-x_1 - x_2 \geq 1$

and $x_1, x_2 \geq 0.$

8. Max. $Z = 3x_1 + 2x_2$

subject to

$2x_1 + x_2 \leq 5$, $x_1 + x_2 \leq 3$

and $x_1, x_2 \geq 0.$

9. Max. $Z = 2x_1 + x_2$

subject to

$x_1 + 2x_2 \leq 10,\ x_1 + x_2 \leq 6$

$x_1 - x_2 \leq 2,\ x_1 - 2x_2 \leq 1$

and $x_1, x_2, x_3 \geq 0$.

10. Min. $Z = 2x_1 + 2x_2$

subject to

$2x_1 + 4x_2 \geq 1,$

$x_1 + 2x_2 \geq 1.$

$2x_1 + x_2 \geq 1,$

and $x_1, x_2 \geq 0$.

11. Max. $Z = 3x_1 + 2x_2$

subject to $x_1 + x_2 \geq 1,$

$x_1 + x_2 \leq 7,$

$x_1 + 2x_2 \leq 10,\ x_2 \leq 3$

and $x_1, x_2 \geq 0$.

12. Formulate the dual of the given L.P.P. and hence solve it:

Min. $Z = 3x_1 - 2x_2 + 4x_3$

subject to

$3x_1 + 5x_2 + 4x_3 \geq 7,$

$6x_1 + x_2 + 3x_3 \geq 4,$

$7x_1 - 2x_2 - x_3 \leq 10,$

$x_1 - 2x_2 + 5x_3 \geq 3$

$4x_1 + 7x_2 - 2x_3 \geq 2$

and $x_1, x_2, x_3 \geq 0$.

13. Solve the following problem by simplex method:

Max. $Z = 30x_1 + 23x_2 + 29x_3$

subject to

$6x_1 + 5x_2 + 3x_3 \leq 26,$

$4x_1 + 2x_2 + 5x_3 \leq 7,$

$x_1, x_2, x_3 \geq 0.$

Also read the solution to the dual of the above problem.

14. Use Dual problem theory to solve the given L.P.P.

Min. $Z = 4x_1 + 3x_2 + 6x_3$

subject to

$x_1 + x_3 \geq 2,$

$x_2 + x_3 \geq 5$

and $x_1, x_2, x_3 \geq 0.$

15. Solve the following L.P.P. using Dual problem theory

Min. $Z = 50x_1 - 80x_2 - 140x_3,$

subject to

$x_1 - x_2 - 3x_3 \geq 4,$

$x_1 - 2x_2 - 2x_3 \geq 3$

and $x_1, x_2, x_3 \geq 0.$

16. Apply the principle of Dual problem to solve the following L.P.P.

Max. $Z = 3x_1 - 2x_2$

subject to

$x_1 + x_2 \leq 5,$

$x_1 \leq 4,\ 1 \leq x_2 \leq 6$

and $x_1, x_2 \geq 0.$

17. Solve the following primal problem and its dual by simplex method

Max. $Z = 5x_1 + 12x_2 + 4x_3$

subject to

$x_1 + 2x_2 + x_3 \leq 5,$

$2x_1 - x_2 + 3x_3 = 2,$

$x_1, x_2, x_3 \geq 0.$

Verify that the solution of the primal can be read from the optimal table of the dual and vice-versa.

18. Use Dual problem theory to solve

Min. $Z = 3x_1 + x_2$

subject to $x_1 + x_2 \geq 1,$

$2x_1 + 3x_2 \geq 2,$

$x_1, x_2, x_3 \geq 0.$

19. Use Dual problem to solve the following L.P.P.

Max. $Z = 4x_1 + 3x_2$

subject to

$x_1 \le 6,\ x_2 \le 8,$

$x_1 + x_2 \le 7,$

$3x_1 + x_2 \le 15,$

$-x_2 \le 1$

and $x_1, x_2 \ge 0.$

20. Write the dual of the following primal

Max. $Z = 40x_1 + 35x_2$

subject to

$2x_1 + 3x_2 \le 60,$

$4x_1 + 3x_2 \le 96,$

$x_1, x_2 \ge 0.$

Solve the primal and the dual by simplex method. Compare the optimal solution of the two problems.

21. Find the dual of the following L.P.P.

Max. $Z = 2x_1 - x_2$

subject to

$x_1 + x_2 \le 10,$

$-2x_1 + x_2 \le 2$

$4x_1 + 3x_2 \ge 12,$

$x_1, x_2, \ge 0.$

Solve the primal problem by simplex method and deduce from it the solution to the dual problem.

22. Find the dual of the following problem

Min. $Z = 6x + 5y + 2x$

subject to $x + 3y + 2z \ge 5,$

$2x + 2y + z \ge 2,$

$4x - 2y + 3z \ge -1.$

$x, y, z \ge 0.$

Hence or otherwise solve the primal problem.

23. Use Dual problem to solve the problem

Min. $Z = 10x_1 + 6x_2 + 2x_3$

subject to

$$-x_1 + x_2 + x_3 \geq 1,$$
$$3x_1 + x_2 - x_3 \geq 2,$$
and $$x_1, x_2, x_3 \geq 0.$$

24. Write the dual of the following problem

Max. $Z = 2x_1 + 3x_2,$

subject to

$$2x_1 + 2x_2 \leq 10,$$
$$2x_1 + x_2 \leq 6,$$
$$x_1 + 2x_2 \leq 6,$$
$$x_1, x_2, \geq 0.$$

Solve the primal and then find the solution to the dual.

25. Max. $Z = x_1 - x_2 + 3x_3$

subject to

$$x_1 + x_2 + x_3 \leq 10$$
$$2x_1 - x_3 \leq 2$$
$$2x_1 - 2x_2 + 3x_3 \leq 6$$
$$x_1, x_2, x_3 \geq 0.$$

26. Max. $Z = 3x_1 + 5x_2 + 4x_3$

subject to

$$2x_1 + 3x_2 \leq 8$$
$$2x_2 + 5x_3 \leq 10$$
$$3x_1 + 2x_2 + 4x_3 \leq 15$$
$$x_1, x_2, x_3 \geq 0.$$

27. Max. $Z = x_1 + 3x_2$

subject to

$$3x_1 + 2x_2 \leq 6$$
$$3x_1 + x_2 = 4,$$
$$x_1, x_2 \geq 0.$$

28. Min. $Z = x_1 + x_2 + x_3$

subject to

$$x_1 - 3x_2 + 4x_3 = 5$$
$$x_1 - 2x_2 \leq 3$$

$2x_2 - x_3 \geq 4.$

$x_1, x_2 \geq 0, x_2$ is unrestricted.

29. Min. $Z = 7x_1 + 3x_2 + 8x_3$

subject to

$8x_1 + 2x_2 + x_3 \geq 3$

$3x_1 + 6x_2 + 4x_3 \geq 4$

$4x_1 + x_2 + 5x_3 \geq 1$

$x_1 + 5x_2 + 2x_3 \geq 7,$

$x_1, x_2, x_3 \geq 0.$

30. Max. $Z = 3x_1 + x_2 + x_3 - x_4$

subject to

$x_1 + 5x_2 + 3x_3 + 4x_4 \leq 5$

$x_1 + x_3 = -1$

$x_3 - x_4 \geq -5,$

$x_1, x_2, x_3, x_4 \geq 0.$

31. Max. $Z = 3x_1 + x_2 + 4x_3 + x_4 + 9x_5$

subject to

$4x_1 - 5x_2 - 9x_3 + x_4 - 2x_5 \leq 6$

$2x_1 + 3x_2 + 4x_3 - 5x_4 + x_5 \leq 9$

$x_1 + x_2 - 5x_3 - 7x_4 + 11x_5 \leq 10,$

$x_1, x_2, x_3, x_4, x_5 \geq 0.$

32. Max. $Z = 3x_1 + x_2 + 2x_3 - x_4$

subject to

$2x_1 - x_2 + 3x_3 + x_4 = 1$

$x_1 + x_2 - x_3 + x_4 = 3,$

$x_1, x_2, x_3 \geq 0, x_4$ is unrestricted in sign.

RESTRICTIONS ON ASSIGNMENT

Sometimes due to some restrictions (technical, legal or others) the assignment of a particular facility to a particular job is not permitted. To overcome such difficulty we assign a very high cost (say, infinite cost) to the corresponding cell, so that the activity will be automatically excluded from the optimal solution.

THE MAXIMAL ASSIGNMENT PROBLEM

The problem of minimizing the total cost. Sometimes the assignment problem deals with the maximization of an objective function rather than to minimize it. For example, the problem may be to assign persons to jobs in such a way that the expected profit is maximum. This problem may be solved easily by first converting it to a minimization problem and then we can use the usual procedure of assignment algorithm. This conversion can be very easily done by modifying the given profit matrix to the cost matrix in either of the two ways.

(i) *Select the greatest element of the given profit matrix and then subtract each element of the matrix from this greatest element to get the modified matrix.*

For, if $[c_{ij}]$ is the given profit matrix and c_{rk} is the greatest element of this matrix then the modified matrix will be $[c_{ij}']$, where $c_{ij}' = c_{rk} - c_{ij}$. It can be shown that if $x_{ij} = x_{ij}$ maximizes $Z = \Sigma\Sigma c_{ij}\, x_{ij}$, then $x_{ij} = X_{ij}$ minimizes the function $Z' = \Sigma\Sigma c_{ij}'\, x_{ij}$, where $c_{ij}' = c_{rk} - c_{ij}$. It follows from the relation

$$Z' = \Sigma\Sigma c_{ij}'\, x_{ij} = \Sigma\Sigma\, (c_{rk} - c_{ij})\, x_{ij} = \Sigma\Sigma c_{rk}\, x_{ij} - \Sigma\Sigma c_{ij}\, x_{ij} = nc_{rk} - Z.$$

(ii) *Place minus sign before each element of the profit matrix to get the modified matrix.*

In this case if $[c_{ij}]$ is the given profit matrix then the modified matrix will be $[c_{ij}']$ where $c_{ij}' = -c_{ij}$. It can be shown that if $x_{ij} = X_{ij}$ maximizes $Z = \Sigma\Sigma c_{ij}\, x_{ij}$ then $x_{ij} = X_{ij}$ minimizes $Z' = \Sigma\Sigma c_{ij}'\, x_{ij}$.

UNBALANCED ASSIGNMENT PROBLEM

In case the number of tasks (jobs) is not equal to the number of facilities (persons), the assignment problem is called an *unbalanced assignment problem.* Thus, the cost matrix of an unbalanced assignment problem is not a square matrix. For the solution of such problem we add the dummy (fictitious) rows or columns to the given matrix with zero costs to form it a square matrix. Then the usual assignment algorithm can be applied to this resulting baseness assignment problem.

Example 1:

One unit of product A contributes Rs. 7 and requires 3 units of raw material and 2 hours of labor. One unit of product B contributes Rs. 5 and requires one unit of raw material and one hour of labour. Availability of the raw material at present is 48 units and there are 40 hours of labor.

(i) *Formulate the linear programming problem,*

(ii) *Write the dual and solve it by simplex method. Also find the optimal product mix.*

Solution:

The linear programming problem corresponding to the given information is

$$\text{Max. } Z = 7x_1 + 5x_2$$

subject to

$$3x_1 + x_2 \leq 48$$
$$2x_1 + x_2 \leq 40$$
$$x_1, x_2 \geq 0.$$

The given L.P.P. is in standard primal form.

The dual problem can be written as

$$\text{Min. } Z_D = 48w_1 + 40w$$

subject to

$$w_1 + 2w_2 \geq 7$$
$$w_1 + w_2 \geq 5,$$
$$w_1, w_2 \geq 0.$$

To solve the dual problem by simplex method, changing the dual objective function to maximization and introducing surplus variables w_3, w_4 and artificial variables w_5, w_6 respectively to change the constraint incredulities into equations. The dual problem can be written as

$$\text{Max. } Z_D' = -48w_1 - 40w_2 + 0w_3 + 0w_4 - Mw_5 - Mw_6$$

subject to

$$3w_1 + 2w_2 - w_3 + w_5 = 7$$
$$w_1 + w_2 - w_4 + w_6 = 5,$$
$$w_1, w_2, \ldots, w_6 \geq 0.$$

The initial B.F.S. is

$$w_1 = 0 = w_2 = w_3 = w_4, w_5 = 7, w_6 = 5.$$

Now applying simplex method to obtain the optimal solution of the dual problem.

		c_j	–48	–40	0	0	–M	–M	Min. ratio
B	c_B	w_B	W_1	W_2	W_3	W_4	W_5	W_6	w_B/W_1
W_5	–M	7	(3)	2	–1	0	1	0	7/3
W_6	–M	5	1	1	0	–1	0	1	5

$Z_D' = -12M$		Δ_j	$4M-48$ ↓	$3M-40$	$-M$	$-M$	0 ↑	0	w_B/W_2
W_1	-48	$7/3$	1	$(2/3)$	$-1/3$	0	$1/3$	0	$7/2$
W_6	$-M$	$8/3$	0	$1/3$	$1/3$	-1	$-1/3$	1	8
$Z_D' = -\left(112+\frac{8}{3}M\right)$		Δ_j	0 ↓	$\frac{M}{3}-8$ ↑	-16	$-M$	$16-\frac{4M}{3}$	0	

		c_j	-48	-40	0	0	$-M$	$-M$	Min. ratio
B	c_B	w_B	W_1	W_2	W_3	W_4	W_5	W_6	w_B/W_1
W_2	-40	$7/2$	$3/2$	1	$-1/2$	0	$-1/2$	0	–
W_6	$-M$	$3/2$	$-1/2$	0	$(1/2)$	-1	$-1/2$	1	3
$Z_D' = -\left(140+\frac{3}{2}M\right)$		Δ_j	$12-\frac{1}{2}M$ ↑	0	$\frac{1}{2}M-20$	$-M$	$20-\frac{3}{2}M$	0 ↓	
W_2	-40	5	1	1	0	-1	0	1	
W_3	0	3	-1	0	1	-2	-1	2	
$Z_D' = -200$		Δ_j	-8	0	0	-40	$-M$	$40-M$	

$\therefore$ Solution of the dual problem is

$$w_1 = 0,\ w_2 = 5,$$

Min. $Z_D = -$ Max. $Z_D' = 200$.

Hence the solution of the primal problem is

$$x_1 = -\Delta_3 = 0,\ x_2 = -\Delta_4 = 40$$

and Max. $Z = 200$.

Example 2:

Write the dual of the following problem

$$\text{Min. } Z = 3x_1 - 2x_2 + 4x_3$$

subject to $3x_1 + 5x_2 + 4x_3 \geq 7$

$$6x_1 + x_2 + 3x_3 \geq 4$$

$$7x_1 - 2x_2 - x_3 \leq 10$$

$$x_1 - 2x_2 + 5x_3 \geq 3$$

$$4x_1 + 7x_2 - 2x_3 \geq 2$$

$$x_1, x_2, x_3 \geq 0.$$

Solution:

First we shall convert the given L.P.P. into standard prima form.

Since the given problem is of minimization therefore all the constraints should have the sign $\geq$.

Thus the standard primal form of the given L.P.P. is

$$\text{Min } Z = 3x_1 - 2x_2 + 4x_3$$

subject to $3x_1 + 5x_2 + 4x_3 \geq 7$

$$6x_1 + x_2 + 3x_3 \geq 4$$

$$-7x_1 + 2x_2 + x_3 \geq -10$$

$$x_1 - 2x_2 + 5x_3 \geq 3$$

$$4x_1 + 7x_2 - 2x_3 \geq 2, x_1,$$

$$x_2, x_3 \geq 0.$$

The matrix form of the above problems

$$\text{Min } Z = (3, -2, 4)(x_1, x_2, x_3)$$

$$\text{subject to} \begin{bmatrix} 3 & 5 & 4 \\ 6 & 1 & 3 \\ -7 & 2 & 1 \\ 1 & -2 & 5 \\ 4 & 7 & -2 \end{bmatrix} \begin{bmatrix} x_1 \\ x_2 \\ x_3 \end{bmatrix} \geq \begin{bmatrix} 7 \\ 4 \\ -10 \\ 3 \\ 2 \end{bmatrix}$$

Now the dual of the given primal is

$$\text{Max. } Z_D = (7, 4, -10, 3, 2)(y_1, y_2, y_3, y_4, y_5)$$

subject to

$$\begin{bmatrix} 3 & 6 & -7 & 1 & 4 \\ 5 & 1 & 2 & -2 & 7 \\ 4 & 3 & 1 & 5 & -2 \end{bmatrix} \begin{bmatrix} y_1 \\ y_2 \\ y_3 \\ y_4 \\ y_5 \end{bmatrix} \leq \begin{bmatrix} 3 \\ -2 \\ 4 \end{bmatrix}$$

or $\quad \text{Max. } Z_D = 7y_1 + 4y_2 - 10y_3 + 3y_4 + 2y_5$

subject to $3y_1 + 6y_2 - 7y_3 + y_4 + 4y_5 \leq 3$

$$5y_1 + y_2 + 2y_3 - 2y_4 + 7y_5 \leq 3$$

$$4y_1 + 3y_2 + y_3 + 5y_4 - 2y_5 \leq 4,$$

$$y_1, y_2, y_3, y_4, y_5 \geq 0.$$

Example 3:

Apply simplex method to solve the following problem

$$Max.\ Z = 30x_1 + 23x_2 + 29x_3$$

subject to $6x_1 = 5x_2 + 3x_3 \leq 26$

$$4x_1 + 2x_2 + 5x_3 \leq 7$$

$$x_1, x_2, x_3 \geq 0.$$

Hence or otherwise find the solution to the dual of the above problem.

Solution:

The given problem is of maximization. Introducing slack variables x_4, x_5 to change the constraint inequalities into equations, we get

$$\text{Max. } Z = 30x_1 + 23x_2 + 29x_3 + 0x_4 + 0x_5,$$

s.t. $\quad 6x_1 + 5x_2 + 3x_3 + x_4 = 26$

$$4x_1 + 2x_2 + 5x_3 + x_5 = 7,$$

$$x_1, x_2, \dots, x_5 \geq 5.$$

The starting B.F.S. is

$$x_1 = 0,\ x_2 = 0,\ x_3 = 0,$$

$$x_4 = 26,\ x_5 = 7.$$

Now applying simplex method to obtain the optimal solution, we have

		c_j	30	23	29	0	0	Min. ratio
B	c_B	x_B	Y_1	Y_2	Y_3	Y_4	Y_5	x_B/Y_1
Y_4	0	26	6	5	3	1	0	26/6
Y_5	0	7	(4)	2	5	0	1	7/4
$Z=c_B$	$x_B=0$	Δ_j	30 ↑	23	29	0	0 ↓	x_B/Y_2
Y_4	0	31/2	0	2	–9/2	1	–3/2	31/4
Y_1	30	7/4	1	(1/2)	5/4	0	1/4	7/2
Z=105/2	Δ_j	0 ↓	8 ↑	–17/2	0	–15/2		
Y_4	0	17/2	–4	0	–19/2	1	–5/2	
Y_2	23	7/2	2	1	5/2	0	1/2	
Z=161/2	Δ_j	–16	0	–57/2	0	–23	/2	

In the last simplex table all $\Delta_j \le 0$ therefore the optimal solution is

$$x_1 = 0,\ x_2 = 7/2,\ x_3 = 0$$

and $\quad$ Max. $Z = 161/2$.

Dual Problem

The given problem is in standard primal form. The dual of the given problem is

$$\text{Min. } Z_D = 26w_1 + 7w_2$$

subject to $6w_1 + 4w_2 \ge 30$

$$5w_1 + 2w_2 \ge 23$$

$$3w_1 + 2w_2 \ge 29,$$

$$w_1, w_2 \ge 0.$$

To read the solution of the dual from the find simplex table of the primal problem.

$$w_1 = -\Delta_4 = 0,$$

$$w_2 = -\Delta_5 = 23/2$$

and $\quad$ min Z_D max $Z = 161/2$.

Example 4:

Write the dual of the following linear programming problem and hence solve it.

$$\text{Max. } Z = 3x_1 - 2x_2$$

subject to $x_1 \le 4$

$$x_2 \le 6$$

$$x_1 + x_2 \le 5$$

$$-x_2 \le -1$$

$$x_1, x_2 \ge 0.$$

Solution:

The given problem is in standard primal form. The dual of the given primal is

$$\text{Min. } Z_D = 4w_1 + 6w_2 + 5w_3 - w_4$$

subject to $w_1 + w_3 \ge 3$

$$w_2 + w_3 - w_4 \ge -2,$$

$$w_1, w_2, w_3, w_4 \ge 0.$$

Changing the dual problem to maximization and introducing surplus variable w_5 and slack variable w_6 to change the inequalities into equations, the dual problem becomes

$$\text{Max.. } Z_D' = -4w_1 - 6w_2 - 5w_3 + w_4 + 0w_5 + 0w_6$$

subject to $w_1 + w_3 - w_5 = 3$

$$-w_2 - w_3 + w_4 + w_6 = 2,$$

$$w_1, w_2, \ldots, w_6 \geq 0.$$

Here, we have not introduced the artificial variable because in the first constraint equation w_1 will serve the purpose.

The starting B.F.S. is

$$w_1 = 3, w_6 = 2.$$

Now applying the simplex method to obtain the optimal solution of the dual problem

		c_j	–4	–6	–5	1	0	0	Min. ratio
B	c_B	w_B	W_1	W_2	W_3	W_4	W_5	W_6	w_B/W_4
W1	–4	3	1	0	1	0	–1	0	–
W_6	0	2	0	–1	–1	(1)	0	1	<u>2</u>
$Z_D' = -12$		Δ_j	0	–6	–1	1 ↑	–4	0 ↓	
W_1	–4	3	1	0	1	0	–1	0	
W_4	1	2	0	–1	–1	1	0	1	
$Z_D'=-10$		Δ_j	0	–5	0	0	–4	–1	

Since all Δ_j are negative or zero therefore the last table gives the optimal solution of the dual.

The optimal solution of the dual is

$$w_1 = 3, w_2 = 0, w_3 = 0, w_4 = 2,$$

$$\text{Min. } Z_D = -\text{ Max. } Z_D' = 10.$$

To read the solution of the primal from the final simplex table of the dual.

The optimal solution of the primal problem is

$$x_1 = -\Delta_5 = 4,\ x_2 = -\Delta_6 = 1 \text{ and Max. } Z = \text{Min. } Z_D = 10.$$

Example 5:

Write the dual of the following problem

$$Max.\ Z = 5x_1 - 2x_2 + 3x_3$$

subject to $2x_1 + 2x_2 - x_3 \geq 2$

$$3x_1 - 4x_2 \leq 2$$

$$x_2 + 2x_3 \leq 5,$$

$$x_1, x_2, x_3 \geq 0.$$

Hence or otherwise write the solution of the primal.

Solution:

Writing the given L.P.P. into standard primal form, we get

$$\text{Max. } Z = 5x_1 - 2x_2 + 3x_3$$

subject to

$$-2x_1 - 2x_2 + x_3 \leq -2$$

$$3x_1 - 4x_2 \leq 3$$

$$x_2 + 2x_3 \leq 5,$$

$$x_1, x_2, x_3 \leq 0.$$

The dual of the given primal is

$$\text{Min. } Z_D = -2w_1 + 3w_2 + 5w_3$$

subject to

$$-2w_1 + 3w_2 \geq 5$$

$$-2w_1 - 4w_2 + w_3 \geq -2$$

$$w_1 + 2w_3 \geq 3,$$

$$w_1, w_2, w_3 \geq 0.$$

Now to solve the dual problem by simplex method, converting it to maximization problem and multiplying both sides of the second constraint by -1, we get

$$\text{Max. } Z_D' = 2w_1 - 3w_2 - 5w_3$$

subject to $-2w_1 + 3w_2 + 0w_3 \geq 5$

$$2w_1 + 4w_2 - w_3 \leq 2$$

$$w_1 + 0w_2 + 3w_3 \geq 3,$$

$w_1, w_2, w_3 \geq 0.$

Introducing surplus variables w_4, w_6, slack variable w_5 and artificial variables A_1, A_2 the above dual problem changes to

$$\text{Max. } Z_D' = 2w_1 - 3w_2 - 5w_3 + 0w_4 + 0w_5 + 0w_6 - MA_1 - MA_2$$

subject to

$$-2w_1 + 3w_2 + 0w_3 - w_4 + A_1 = 5$$

$$2w_1 + 4w_2 - w_3 + 0w_4 + w_5 = 2$$

$$w_1 + 0w_2 + 3w_3 - w_6 + A_2 = 3.$$

$$w_1, w_2, \ldots, w_6 \geq 0.$$

The starting B.F.S. is $w_1 = 0 = w_2 = w_3 = w_4 = w_6$, $w_5 = 2$, $A_1 = 5$, $A_2 = 3$.

Now apply the simplex method to find the optimal solution of dual problem.

		c_j	2	−3	−5	0	0	0	−M	−M	Min. ratio
B	c_B	x_B	W_1	W_2	W_3	W_4	W_5	W_6	A_1	A_2	w_B/W_2
A_1	−M	5	−2	3	0	−1	0	0	1	0	5/3
W_5	0	2	2	(4)	−1	0	1	0	0	0	2/4
A_2	−M	3	1	0	3	0	0	−1	0	1	–
$Z_D' = -8M$		Δ_j	2 − M	− 3 + 3M	− 5 + 3M		−M	0	−M	0	0 w_B/W_3
A_1	−M	7/2	−7/2	0	3/4	−1	−3/4	0	1	0	14/3
W_2	−3	1/2	1/2	1	−1/4	0	1/4	0	0	0	–
A_2	−M	3	1	0	(3)	0	0	−1	0	1	1
$Z_D' = -\,\Delta_j$ $\frac{13}{2}M - \frac{3}{2}$			$\frac{7-5M}{2}$	0 ↑	$\frac{15M-23}{4}$	−M	$\frac{303M}{4}$	−M	0 ↓	0	w_B/W_6
A_1	−M	11/4	−15/4	0	0	−1	−3/4	(1/4)	1	11	
W_2	−3	3/4	7/12	1	0	0	1/4	−1/12	0	–	
W_3	−5	1	1/3	0	1	0	0	−1/3	0	–	
$Z_D' = -\frac{11}{4}M - \frac{29}{4}$			Δ_j	$\frac{65-45M}{12}$	0	0	−M	$\frac{3-3M}{4}$	$\frac{3M-23}{12}$ ↓	0 ↓	

W_6	0	11	–15	0	0	–4	–3	1
W_2	–3	5/3	–2/3	1	0	–1/3	0	0
W_3	–5	14/3	–14/3	0	1	–4/3	–1	0
$Z_D' = -\frac{85}{3}$	Δ_j	–70/3	0	0	–23/3	–5	0	

In the last simplex table all $\Delta_j \leq 0$ therefore it gives optimal solution of the dual problem.

The optimal solution of the dual is

$$w_1 = 0,\ w_2 = 5/3,\ w_3 = 14/3$$

and $\quad$ Min. $Z_D = -$ Max. $Z_D' = 85/3$.

Now the optimal solution of the primal is

$$x_1 = -\Delta_4 = 23/3,\ x_2 = -\Delta_5 = 5,$$

$$x_3 = -\Delta_6 = 0$$

and $\quad$ Max. $Z =$ Min. $Z_D = 85/3$.

Example 6:

Use Dual problem to solve

Min. $Z = 3x_1 + x_2$

subject to $x_1 + x_2 \geq 1$

$$2x_1 + 3x_2 \geq 2,$$

$$x_1, x_2 \geq 0.$$

Solution:

The given L.P.P. is in standard primal form. The dual problem is given by

Max. $Z_D = w_1 + 2w_2$

subject to

$$w_1 + 2w_2 \leq 3$$

$$w_1 + 3w_2 \leq 1,\ w_1,\ w_2 \geq 0.$$

Introducing slack variables w_3 and w_4 to change the constraint inequalities into equations, the dual problem becomes

Max. $Z_D = w_1 + 2w_2 + 0w_3 + 0w_4$

subject to

$$w_1 + 2w_2 + w_3 = 3$$

$$w_1 + 3w_2 + w_4 = 1,$$

$w_1, w_2, w_3, w_4 \geq 0.$

Initial B.F.S. is $w_3 = 3$, $w_4 = 1$.

Now we shall apply simplex method to obtain solution.

		c_j	1	2	0	0	Min ratio
B	c_B	w_B	W_1	W_2	W_3	W_4	w_B/W_2
W_3	0	3	1	2	1	0	3/2
W_4	0	1	1	(3)	0	1	1/3
$Z_D = 0$		Δ_j	1	2 ↑	0	0 ↓	w_B/W_1
W_3	0	7/3	1/3	0	1	–2/3	7
W_2	2	1/3	(1/3)	1	0	1/3	1
$Z_D = 2/3$		Δ_j	1/3 ↑	0 ↓	0	–2/3	
W_3	0	2	0	–1	1	–1	
W_1	1	1	1	3	0	1	
$Z_D = 1$	Δ_j	0	–1	0	–1		

Since all $\Delta_j \leq 0$, therefore the dual problem has optimal solution

$w_1 = 1, w_2 = 0,$

Max. $Z_D = 1.$

Now the solution of the original primal problem form the last simplex table of the dual is

$x_1 = -\Delta_3 = 0,$

$x_2 = -\Delta_4 = 1,$

min. Z = max. $Z_D = 1.$

Thus, the optimal product mix is: none of the product A and 40 units of product B for a total conscription of Rs. 200.

Example 7:

In a machine shop a supervisor wishes to assign five jobs among six machines any one of the job can be processed completely by any one of the machines the cost (in Rs.) of processing any jobs on any machine is given below.

Jobs

	A	*B*	*C*	*D*	*E*	*F*
1	*13*	*13*	*16*	*23*	*19*	*9*
2	*11*	*19*	*26*	*16*	*17*	*18*
3	*12*	*11*	*4*	*9*	*6*	*10*
4	*7*	*15*	*9*	*14*	*14*	*13*
5	*9*	*13*	*12*	*8*	*14*	*11*

The assignment of jobs to machines must be one-to-one basis. Assign the jobs to machines so that the total cost is minimum. Find the minimum total cost.

Solution:

We add a dummy sixth row (job 6) in the cost matrix so as to get the following balanced assignment problem:

	A	*B*	*C*	*D*	*E*	*F*
1	13	13	16	23	19	9
2	11	19	26	16	17	18
3	12	11	4	9	6	10
4	7	15	9	14	14	13
5	9	13	12	8	14	11
6	0	0	0	0	0	0

Now following the usual procedure of solving an assignment problem, an optimum assignment is obtained as follows:

1 → F, 2 → A, 3 → E, 4 → C, 5 → D

with total minimum cost as

Rs. (9 + 11 + 6 + 9 + 8), *i.e.*, Rs. 43.

A company has 4 machines on which to do 3 jobs. Each job can be assigned to one and only one machine. The cost of each job on each machine is given in the following table:

Machine

Job	*W*	*X*	*Y*	*Z*
A	18	24	28	32
B	8	13	17	19
C	10	15	19	22

What are the job assignments which will minimize the cost?

A department head has six jobs and five subordinates. The subordinates differ in their efficiency and the tasks differ. In third intrinsic difficulty. The department head estimates the time each man would take to perform each task as given in the effectiveness matrix below:

Man	*Task*					
	A	*B*	*C*	*D*	*E*	*F*
1	20	15	26	40	32	12
2	15	32	46	26	28	20
3	11	15	2	12	6	14
4	8	24	12	22	22	20
5	12	20	18	10	22	15

Only one task can be assigned to one man determine how should the job be allocated so as to minimize the total man hour. Find the minimum total man hours.

Example 8:

A company has four territories open and four salesmen available for assignment. The territories are not equally rich in their sales potential; is estimated that a typical salesman operating in each territory would bring in the following annual sales:

Territory	:	*I*	*II*	*III*	*IV*
Annual sales (Rs)	:	*60,000*	*50,000*	*40,000*	*30,000*

Four salesmen are also considered to differ in their ability : it is estimated that, working under the same conditions, their yearly sales would be proportionately as follows:

Salesman	:	*A*	*B*	*C*	*D*
Proportion	:	*7*	*5*	*5*	*4*

If the criterion is maximum expected total sales, then intuitive answer is to assign the best salesman to the richest territory, the next best salesman to the second richest, and so on. Verify this answer by the assignment technique.

Solution:

To construct the effectiveness matrix.

The sum of proportions of sales of four salesmen

$$= 7 + 5 + 5 + 4 = 21.$$

If we consider Rs 10,000 as one unit, then A's annual sales in the four territories are

$$\frac{7}{21}\times 6, \frac{7}{21}\times 5, \frac{7}{21}\times 4, \frac{7}{21}\times 3.$$

Similarly we can calculate the annual sales of other salesmen in different territories. In order to avoid fractional values we consider the sales in 21 years.

Thus, the maximum sale matrix is obtained as follows:

Sales in Rs 10,000 →		*6*	*5*	*4*	*3*
Sales proportion ↓		*I*	*II*	*III*	*IV*
7	A	42	35	28	21
5	B	30	25	20	15
5	C	30	25	20	15
4	D	24	20	16	12

This is a 'maximization' problem. To convert it into a 'minimization' one let us multiply each element of the above matrix by – 1. Thus resulting matrix becomes:

	I	*II*	*III*	*IV*
A	–42	–35	–28	–21
B	–30	–25	–20	–15
C	–30	–25	–20	–15
D	–24	–20	–16	–12

Now we shall solve the minimization problem by usual assignment algorithm.

Step 1. Subtracting the smallest element of each row from every element of that row and then subtracting the smallest element of each column from every element of that column the reduced matrix is shown in the following table.

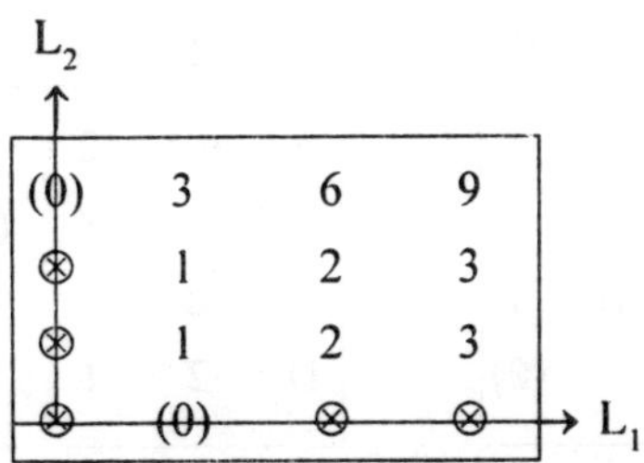

Step 2. Make zero assignments. Since total no. of marked '□' zeros is less than 4, so an optimal assignment is not possible at this stage (as shown in the table of step 1).

Step 3. We draw minimum no. of lines to cover all the zeros.

The minimum element among all uncovered elements is 1.

Then subtract this value 1 from each uncovered element, add 1 to the elements which lie at the intersection of two lines L_1, L_2 and leave other elements unchanged. Thus, the revised matrix is shown in the table.

	L_2	L_3			
	(0)	2	5	8	√ (4)
	⊗	(0)	1	2	√ (5)
	⊗	⊗	1	2	√ (1)
L1	1	⊗	(0)	⊗	
	√ (2)	√ (3)			

Step 4. Giving the 'zero assignments' in the table of step 3, we see that there is no assignment in row 3 and column 4. So, we again proceed to draw minimum number of lines to cover all zeros at least once.

Step 5. We find that three lines L_1, L_2, L_3 cover all the zeros. Again minimum element among all uncovered element is 1. Subtracting 1 from all the uncovered elements, adding 1 to the elements which lie at the intersection of the lines L_1, L_2, L_3 and leaving other elements unchanged, we get the following matrix:

0	2	4	7
0	0	0	1
0	0	0	1
2	1	0	0

Step 6. Giving 'zero assignments', we get two assignments as shown in the following tables:

	I	II	III	IV
A	(0)	2	4	7
B	⊗	(0)	⊗	1
C	⊗	⊗	(0)	1
D	2	1	⊗	(0)

	I	II	III	IV
A	(0)	2	4	7
B	⊗	⊗	(0)	1
C	⊗	(0)	⊗	1
D	2	1	⊗	(0)

Thus two optimal solution are

(i) A → I, B → II, C → III, D → IV

(i) A → I, B → III, C → II, D → IV.

From both the solution, it is obvious that the best salesman A is assigned to the richest territory I, the worst salesman D to the poorest territory IV. Salesman B and C being equally good.

Example 9:

A company has 5 jobs to be done. The following matrix shows the return in rupees on assigning ith (i = 1, 2, 3, 4, 5) machine to the jth job (j = A, B, C, D, E). Assign the five jobs to the five machines so as to maximize the total expected profit.

		Jobs				
		A	*B*	*C*	*D*	*E*
	1	*5*	*11*	*10*	*12*	*4*
	2	*2*	*4*	*6*	*3*	*5*
Machines	*3*	*3*	*12*	*5*	*14*	*6*
	4	*6*	*14*	*4*	*11*	*7*
	5	*7*	*9*	*8*	*12*	*5*

Solution:

First we shall convect the problem from maximization to minimization. The greatest element of the given matrix is 14. Subtracting all the elements of the given matrix from 14, the modified matrix is obtained. Now we shall follow the usual procedure of solving an assignment problem.

9	3	4	2	10
12	10	8	11	9
11	2	9	0	8
8	0	10	3	7
7	5	6	2	9

Step 1. Subtracting the minimum element of each row from every element of the corresponding row and then subtracting the minimum element of each column from every element of the corresponding column, the modified matrix reduces to

3	1	2	0	7
0	2	0	3	0
7	2	9	0	7
4	0	10	3	6
1	3	4	0	6

Step 2. Now make 'zero assignments'. The following tables show the necessary steps for reaching the optimal assignment : 1 → C, 2 → E, 3 → D, 4 → B, 5 → A with maximum profit of Rs. 50.

3	1	2	(0)	7
(0)	2	⊗	3	⊗
7	2	9	⊗	7
4	(0)	10	3	6
1	3	4	⊗	6

2	⊗	1	⊗	6
⊗	2	(0)	4	⊗
6	1	8	(0)	6
4	(0)	10	4	6
(0)	2	3	⊗	5

1	⊗	(0)	⊗	5
⊗	3	⊗	5	(0)
5	1	7	(0)	5
3	(0)	9	4	5
(0)	3	3	1	5

Example 10:

Four engineers are available to design four projects. Engineer 2 is not competent to design the project B. Given the following time estimates needed to each engineer to design a given project, find how should the engineers be assigned to projects so as to minimize the total design time of four projects.

		Projects			
		A	*B*	*C*	*D*
Engineers	*1*	*12*	*10*	*10*	*8*
	2	*14*	*Not suitable*	*15*	*11*
	3	*6*	*10*	*16*	*4*
	4	*8*	*10*	*9*	*7*

Solution:

To avoid the assignment 2 → B, we take its time to be very large (say ∞). Then the cost matrix of the resulting assignment problem is shown in the following table:

12	10	10	8
14	∞	15	11
6	10	16	4
8	10	9	7

Now we apply the assignment technique in the usual manner. The following tables show the necessary steps for reaching the optimal solution:

4	2	2	0
3	∞	4	0
2	6	12	(0)
1	3	2	0

3	(0)	⊗	⊗
2	∞	2	(0)
1	4	10	⊗
(0)	1	⊗	⊗

3	(0)	⊗	1
1	∞	1	(0)
(0)	3	9	⊗
⊗	1	(0)	1

Thus the optimal assignment is : 1 → B, 2 → D, 3 → A, 4 → C.

Total minimum cost (time) = 10 + 11 + 6 + 9 = 36.

Example 11:

A job shop has purchased 5 new machines of different types. Theorem 5 available locations in the shop where a machine could be installed. Some of these locations are more desirable than others for particular machines because of their proximity to work centres which would have a heavy work flow to and from these machines. Therefore, the objective is to assign the new machines to the available locations in order to minimize the total cost of material handling. The estimated cost per unit time of materials handling involving each of the machines is given below for the respective locations. Locations 1, 2, 3, 4 and 5 are not considered suitable for machines A, B, C, D and E respectively. Find the optimal solution:

Location (cost in Rs.)

Machines	*1*	*2*	*3*	*4*	*5*
A	×	*10*	*25*	*25*	*10*
B	*1*	×	*10*	*15*	*2*
C	*8*	*9*	×	*20*	*10*
D	*14*	*10*	*24*	×	*15*
E	*10*	*8*	*25*	*27*	×

How will the optimal solution get modified if location 5 is also unsuitable for machine A?

Solution:

Since locations 1, 2, 3, 4 and 5 are not suitable for machines A, B, C, D and E respectively, an extremely large cost (say ∞) should be attached to these locations. The resulting assignment problem cost matrix is shown below:

∞	10	25	25	10
1	∞	10	15	2
8	9	∞	20	10
14	10	24	∞	15
10	8	25	27	∞

∞	2	6	3	0
⊗	∞	0	2	1
⊗	3	∞	0	2
2	0	3	∞	3
⊗	0	6	5	∞

Now, following the usual procedure of solving an assignment problem, an optimum assignment is obtained as displayed above. This gives us the optimum assignment as

A → 5, B → 3, C → 4, D → 2, and E → 1

with total minimum cost as Rs. (10 + 10 + 20 + 10 + 10), *i.e.,* Rs. 60.

Now, if location 5 is also unsuitable for machine A, we attach an extremely large cost (= ∞) to cell (1, 5). Now applying the assignment algorithm to this modified problem, the following assignment algorithm can easily be obtained:

A → 4, B → 3, C → 5,

D → 2, and E → 1

or A → 2, B → 3, C → 4,

D → 5, and E → 1

with minimum total cost of Rs. 65.

Example 12:

A company is faced with the problem of assigning six different machines to five different jobs. The costs are estimated as follows (in hundreds of rupees):

		Jobs				
		1	*2*	*3*	*4*	*5*
	1	*2.5*	*5*	*1*	*6*	*1*
	2	*2*	*5*	*1.5*	*7*	*3*
Machines	*3*	*3*	*6.5*	*2*	*8*	*3*
	4	*3.5*	*7*	*2*	*9*	*4.5*
	5	*4*	*7*	*3*	*9*	*6*
	6	*6*	*9*	*5*	*10*	*6*

Solve the problem assuming that the objective is to minimize the total cost.

Solution:

Since the matrix is not square, it is an unbalanced assignment problem. We introduce one fictitious job (6th column with zero costs) to get a square matrix. Further, the matrix involves elements is decimal. We can make then complete numbers 'by multiplying each element of the cost matrix by 2. The new modified cost matrix is shown in the following table.

	1	2	3	4	5	6
1	5	10	2	12	2	0
2	4	10	3	14	6	0
3	6	13	4	16	6	0
4	7	14	4	18	9	0
5	8	14	6	18	12	0
6	12	18	10	20	12	0

Applying the usual procedure, finally we get the following table:

	1	2	3	4	5	6
1	2	0	0	0	0	4
2	0	0	1	2	2	4
3	0	1	0	2	0	2
4	1	2	0	4	3	2
5	0	0	0	2	4	0
6	4	4	4	4	4	0

Giving the zero assignments, we get the following tables:

	1	2	3	4	5	6
1	2	⊗	⊗	(0)	⊗	4
2	(0)	⊗	1	2	2	4
3	⊗	1	⊗	2	(0)	2
4	1	2	(0)	4	3	2
5	⊗	(0)	⊗	2	4	⊗
6	4	4	4	4	4	(0)

	1	2	3	4	5	6
1	2	⊗	⊗	(0)	⊗	4
2	⊗	(0)	1	2	2	4
3	⊗	1	⊗	2	(0)	2
4	1	2	(0)	4	3	2
5	(0)	⊗	⊗	2	4	⊗
6	4	4	4	4	4	(0)

Thus the original problem has two alternative optimum solution:

(i) 1 → 4, 2 → 1, 3 → 5, 4 → 3, 5 → 2

(ii) 1 → 4, 2 → 2, 3 → 5, 4 → 3, 5 → 1.

The 6th machine will be left unassigned.

From the original matrix, the total minimum cost = 20 *i.e.,* Rs. 2000 (in each case).

Example 13:

A department head has four tasks to be performed and three subordinates. The subordinates differ in efficiency. The estimates of the time, each subordinate would take to perform, are given below in the matrix. How should he allocate the tasks, one to each man, so as to minimize the total man hours?

		Subordinates		
		1	*2*	*3*
	I	*9*	*26*	*15*
Tasks	*II*	*13*	*27*	*6*
	III	*35*	*20*	*15*
	IV	*18*	*30*	*20*

Solution:

Since the matrix is not square, it is an unbalanced assignment problem. We introduce one fictitious subordinate (4th column with zero costs) to get a square matrix. Thus the resulting matrix is shown in the following table. Now the problem can be solved by usual method.

	1	2	3	4
I	9	26	15	0
II	13	27	6	0
III	35	20	15	0
IV	18	30	20	0

Step 1. Subtracting the minimum element of each row from every element of the corresponding row and then subtracting the minimum element of each column from every element of the corresponding column, the matrix reduces to

	1	*2*	*3*	*4*
I	0	6	9	0
II	4	7	0	0
III	26	0	9	0
IV	9	10	14	0

Step 2. Make the 'zero assignments' in usual manner. The illustration is shown in the following table.

	1	*2*	*3*	*4*
I	(0)	6	9	⊗
II	4	7	(0)	⊗
III	26	(0)	9	⊗
IV	9	10	14	(0)

Since in the table every row and every column have one assignment, so we have the complete optimal zero assignment as

I → 1, II → 3, III → 2. Take IV will be left unassigned.

From the original matrix, the total time (man hours) = 9 + 6 + 20 = 35.

Example 14:

Give the dual of the following problem and solve

Mini. $Z_P = x_1 + x_2,\ 2x_1 + x_2 \geq 4,\ x_1 + 7x_2 \geq 7,\ x_1, x_2 \geq 0.$

Solution:

The dual of the given problem is given by

Max. $Z_D = 4w_1 + 7w_2$

s.t. $2w_1 + w_2 \leq 1$

$w_1 + 7w_2 \leq 1$

$w_1, w_2 \geq 0$

Introducing the slack variable w_{1s} and w_{2s} respectively. The dual problem can be written as

Max. $Z_D = 4w_1 + 7w_2$

s.t. $2w_1 + w_2 + w_{1s} = 1$

$w_1 + 7w_2 + w_{2s} = 1$

$w_1, w_2, w_{1s}, w_{2s} \geq 0$

Taking $w_1 = 0 = w_2$, we have $w_{1s} = 1$, $w_{2s} = 1$, which is the starting basic feasible solution of the dual.

First Table

		c_J	4	7	0	0	$\frac{w_B}{y_{i2}}$
B	c_B	w_B	Y_1	Y_2	Y_{1s}	Y_{2s}	
x_{1s}	0	4	2	1	1	0	4
x_{2s}	0	7	1	7	0	1	1 mini
		$Z_J - c_J$	–4	–7	0	0	

$\Delta_1 = -4$, $\Delta_2 = -7$

Since $\Delta_2 = -7$ is mini of Δ_1 and Δ_2 so y_2 is incoming vector and by minimum ratio rule outgoing vector is y_{2s}.

The key element is 7.

Proceeding as before the second simplex table.

Second Table

		c_J	4	7	0	0	$\frac{w_B}{Y_{i1}}$
B	c_B	w_B	Y_1	y_2	Y_{1s}	Y_{2s}	
x_{1s}	0	3	13/7	0	1	–1/7	21/13 mini
x_2	7	1	1/7	1	0	1/7	7
		$Z_J - c_J$	–3	0	0	1	
			↑		↓		

Since $\Delta_1 = -3$ is mini so y_1 is incoming vector and by mini rule method outgoing vector is y_{1s}.

The key element is $y_{11} = 13/7$

Proceeding as before the third simplex table.

Third Table

		c_J	4	7	0	0
B	c_B	w_B	Y_1	Y_2	Y_{1s}	Y_{2s}
x_1	4	21/13	1	0	7/13	–1/13
x_2	7	10/13	0	1	–1/13	1/13
		$Z_J - c_J$	0	0	$\frac{21}{13}$	10/13

Since all $Z_J - c_J \geq 0$, so this solution of the dual problem is optimal solution of the dual is

$w_1 = 21/13$, $w_2 = 10/13$; Max $Z_D = \frac{31}{13}$ solution of the primal problem is $x_1 = \frac{21}{13}$, $x_2 = 10/13$; Mini. $Z_P = 31/13$.

EXERCISES

1. What is a shadow price? How does the concept relate to the dual of an LP problem?
2. State and prove the relationship between the feasible solutions of LP problem and its dual.
3. Prove that the necessary and sufficient condition for any LP problem and its dual to have optimal solutions is that both have feasible solutions.
4. How the concept of duality can be useful in managerial decision-making?
5. Find the solution of the following problem by simplex method:

   Max. $Z_x = 40x_1 + 50x_2$

   subject to $2x_1 + 3x_2 \leq 3$, $8x_1 + 4x_2 \leq 5$, $x_1, x_2 \geq 0$.
6. A national truck-rental service has a surplus of one truck in each of the cities 1, 2, 3, 4, 5, 6 and a deficit of one truck in each of the cities 7, 8, 9, 10, 11, 12. The distances (in kilometer) between the cities with a surplus and the cities with a deficit are displayed below:

		To					
		7	*8*	*9*	*10*	*11*	*12*
	1	31	62	29	42	15	41
	2	12	19	39	55	71	40
	3	17	29	50	41	22	22
From	4	35	40	38	42	27	33
	5	19	30	29	16	20	23
	6	72	30	30	50	41	20

   How should the trucks be dispersed so as to minimize the total distance travelled?
7. Define the dual of a linear programming problem. State the functional properties of duality.
8. Explain the primal-dual relationship.

9. Discuss in brief 'duality' in linear programming.

Use principle of Dual problem to solve the following problems:

10. A company makes three products X, Y, Z out of three raw materials A, B, C. The number of units of raw materials required to produce one unit of the product is as given in the following table:

	X	*Y*	*Z*
A	1	2	1
B	2	1	4
C	2	5	1

The unit profit contribution of the products X, Y and Z are Rs. 40, 25 and 50 respectively.

The number of units of raw materials available are 36, 60 and 45 respectively.

(i) Determine the product mix that will maximize the total profit.

(ii) Through the final simplex table write the solution to the dual problem.

11. One unit of product A contributes Rs. 7 and requires 3 units of raw material and 2 hours of labour. One unit of product B contributes Rs. 5 and requires one unit of raw material and one hour of labour. Availability of raw material at present is 48 units and there are 40 hours of labour,

(a) Formulate this problem as a linear programming problem.

(b) Write its dual.

(c) Solve the dual by the simplex method and find the optimal product mix and the shadow prices of the raw material and labour.

12. What is the principle of duality in linear programming? Explain its advantages.

13. What is duality? What is the significance of dual variables in a LP model?

14. State the general rules for formulating a dual LP problem from its primal.

15. A company wishes to get at least 160 million 'audience exposures' everytime one of the advertisements is seen or heard by a person. Because of the nature of the product, the company wants at least 60 million of these exposures to involve persons with family income of

over Rs 10,000 a year and at least 80 million of the exposures to involve person between 18 and 40 years of age. The relevant information pertaining to the two advertising media under considerationùmagazine and television, is given below.

	Magazine	*Television*
Cost per advertisement (Rs thousand)	40	200
Audience per advertisement (million)	4	40
Audience per advertisement with income over Rs 10,000 (million)	3	10
Audience (per advertisement) in the age group 18-40 (million)	8	10

The company wishes to determine the number of advertisements each to be released in magazines and television so as to keep the advertisement expenditure to the minimum.

Formulate this problem as a linear programming problem. What will be the minimum expenditure and its allocation among the two media? Write the 'dual' of this problem. You may solve the 'dual' problem to find answer to the problem on hand.

16. Find the optimal assignment for the problems with given cost matrix

(i)

	I	*II*	*III*	*IV*
A	5	3	1	8
B	7	9	2	6
C	6	4	5	7
D	5	7	7	6

(ii)

	1	*2*	*3*	*4*
A	10	12	19	11
B	5	10	7	8
C	12	14	13	11
D	8	15	11	9

17. A diet conscious housewife wishes to ensure certain minimum intake of vitamins A, B and C for the family. The minimum daily (quantity) needs of the vitamins A, B and C for the family are respectively 30, 20 and 16 units. For the supply of these minimum vitamin requirements,

the housewife relies on two fresh foods I and II. The first one provides 7, 5 and 2 units of the three vitamins per gram, respectively and the second one provides 2, 4 and 8 units of the same three vitamins per gram of the foodstuff, respectively. The first foodstuff costs Rs. 3 per gram and the second Rs. 2 per gram. The problem is how many grams of each foodstuff should the housewife buy everyday to keep her food bill as low as possible?

(a) Formulate this problem as an LP model.

(b) Write and then solve the dual problem.

18. The XYZ Manufacturing Company operates a three-shift system at one of its plants. In a certain section of the plant, the number of operators required on each of the three shifts is as follows: 194 Operations Research: Theory and Applications

	Shif	*Number of Operators*
Day	6 A.M.-2 P.M.	50
Afternoon	2 P.M.-10 P.M.	24
Night	10 P.M.-6 A.M.	10

The company pays its operators at the basic rate of Rs. 10 per hour for those working on the day shift. For the afternoon and night shifts, the rates are one-and-a-half times the basic rate and twice the basic rate, respectively. In agreement with each operator at the commencement of his employment, he is allocated to one of these schemes A, B or C. These are as follows:

A : Work (on average) one night shift, one afternoon shift, and two day shifts in every four shifts.

B : Work (on average) equal number of day and afternoon shifts.

C : Work day shifts only.

In schemes A and B, it is necessary to work strictly alternating sequences of specified shifts, as long as the correct proportion of shifts is worked in the long run.

(a) Formulate an LP model to obtain the required number of operators at minimum cost.

(b) By solving the dual of the problem, determine how many operators must be employed under each of the three schemes. Does this result in over-provision of operators on any one of the three shifts?

19. What is an Assignment problem? Give two areas of its applications.
20. If (x_{ij}), i = 1, 2, ... ,n; j = 1, 2, ... , n is an optimal solution for an assignment problem with cost (c_{ij}), then it is also optimal for the problem with cost (c_{ij}') when

$$c_{ij}' = c_{ij} \text{ for } i, j = 1, 2, \ldots, n; j \neq k$$
$$c_{ik}' = c_{ik} - A, \text{ where A is a constant.}$$

21. A computer centre has got three expert programmers. The center needs three application programmed to be developed. The head of the computer centre, after studying carefully the programmed to be developed. Estimates the computer time in minutes required by the experts to the application programmes as follows:

		Programme		
		A	*B*	*C*
	1	120	100	80
Programmer	2	70	90	110
	3	110	140	120

Assign the programmers to the programmes in such a way that total computer time is least.

22. Find the optimal solution for the assignment problem with the following cost matrix:

	I	*II*	*III*	*IV*	*V*
A	11	17	8	16	20
B	9	7	12	6	15
C	13	16	15	12	16
D	21	24	17	28	26
E	14	10	12	11	15

23. A medical scientist claims to have found a cure for the common cold that consists of three drugs called K, S and H. His results indicate that the minimum daily adult dosage for effective treatment is 10 mg of drug K, 6 mg of drug S and 8 mg of drug H. Two substances are readily available for preparing pills for distribution to cold sufferers. Both substances contain all three of the required drugs. Each unit of substance

A contains 6 mg, 1 mg and 2 mg of drugs K, S and H respectively and each unit of substance B contains 2 mg, 3 mg and 2 mg of the same drugs. Substance A costs Rs. 3 per unit and substance B costs Rs. 5 per unit.

(a) Find the least-cost combination of the two substances that will yield a pill designed to contain the minimum daily recommended adult dosage.

(b) Suppose that the costs of the two substances are interchanged so that substance A costs Rs. 5 per unit and substance B costs Rs. 3 per unit. Find the new optimal solution.

24. Write the dual of the given L.P.P. Also find the solution of the dual problem.

25. Solve the following minimal assignment problems:

(i)

	Jobs				
	1	*2*	*3*	*4*	*5*
A	8	4	2	6	1
B	0	9	5	5	4
			Persons		
C	3	8	9	2	6
D	4	3	1	0	3
E	9	5	8	9	5

(ii)

		Man				
		1	*2*	*3*	*4*	*5*
	I	12	8	7	15	4
	II	7	9	17	14	10
	III	9	6	12	6	7
Job	IV	7	6	14	6	10
	V	9	6	12	10	6

(iii)

	Jobs				
	I	*II*	*III*	*IV*	*V*
A	5	11	10	12	4
B	2	4	6	3	5

	Machines				
C	3	12	5	14	6
D	6	14	4	11	7
E	7	9	8	12	5

(iv)

Man

		I	*II*	*III*	*IV*	*V*
	A	1	3	2	3	6
	B	2	4	3	1	5
Job	C	5	6	3	4	6
	D	3	1	4	2	2
	E	1	5	6	5	4

26. One car is available at each of the stations 1, 2, 3, 4, 5, 6 and one car is required at each of the stations 7, 8, 9, 10, 11, 12. The distances between the various stations are given in the matrix below. How should the cars be despatched so as to minimize the total mileage covered?

	7	*8*	*9*	*10*	*11*	*12*
1	41	72	39	52	25	51
2	22	29	49	65	81	50
3	27	39	60	51	32	32
4	45	50	48	52	37	43
5	29	40	39	26	30	33
6	82	40	40	60	51	30

27. The owner of a small machine shop has four persons variable to assign to jobs for the day. Five jobs are offered with the expected profit in rupees for each person on each job being as follows:

Job

	A	*B*	*C*	*D*	*E*
1	6.20	7.80	5.00	10.10	8.20
2	7.10	8.40	6.10	7.30	5.90
Persons 3	8.70	9.20	11.10	7.10	8.10
4	4.80	6.40	8.70	7.70	8.00

Find the assignment of persons to jobs that will result in a maximum profit. Which job shadow declined?

28. Five wagons are available at five stations 1, 2, 3, 4, and 5. These are required at five stations I, II, III, IV and V. The mileages between various stations are given by the following table:

29. Five operators have to be assigned to five machines. The assignment costs are given in the table below:

Machines

		I	*II*	*III*	*IV*	*V*
	A	5	5	–	2	6
	B	7	4	2	3	4
Operators	C	9	3	5	–	3
	D	7	2	6	7	2
	E	6	5	7	9	1

Operator A cannot operate machine III and operator C cannot operate machine IV. Find the optimal assignment schedule.

30. There are 3 persons P_1, P_2 and P_3 and 5 jobs J_1, J_2, ... J_5. Each person can do only one job and a job is to be done by one person only. Using Hungarian method, find which 2 jobs should be left undone in the following cost minimizing assignment problem.

	J_1	J_2	J_3	J_4	J_5
P_1	7	8	6	5	9
P_2	9	6	7	6	10
P_3	8	7	9	5	6

31. A company produces three products: P, Q and R from three raw materials A, B and C. One unit of product P requires 2 units of A and 3 units of B. A unit of product Q requires 2 units of B and 5 units of C and one unit of product R requires 3 units of A, 2 units of B and 4 units of C. The company has 8 units of material A, 10 units of material B and 15 units of material C available to it. Profits per unit of products P, Q and R are Rs. 3, Rs. 5 and Rs. 4, respectively.
   (a) Formulate this problem as an LP problem.
   (b) How many units of each product should be produced to maximize profit?

(c) Write the dual of this problem.

32. Use the Hungarian method to find which of the two jobs should be left undone when each of the 4 persons will do only one job in the following cost minimizing assignment problem:

Job

		J_1	J_2	J_3	J_4	J_5	J_6
	P_1	10	9	11	12	8	5
	P_2	12	10	9	11	9	4
Persons	P_3	8	11	10	7	12	4
	P_4	10	7	8	10	10	5

33. The jobs A, B, C are to be assigned to three machines X, Y, Z. The processing costs (Rs.) are as given in the matrix shown below. Find the allocation which will minimize the overall processing cost.

Machines

		X	*Y*	*Z*
	A	19	28	31
Jobs	B	11	17	16
	C	12	15	13

34. Use Hungarian method to solve the following cost minimizing assignment problem:

	1	*2*	*3*	*4*
I	20	22	28	15
II	16	20	12	13
III	19	23	14	25
IV	10	16	12	10

35. Give the mathematical formulation of an assignment problem.

36. Give an algorithm to solve an assignment problem.

37. Describe the Hungarian method of solving the assignment problem.

38. Three food products are available at costs of Rs. 10, Rs. 36 and Rs. 24 per unit, respectively. They contain 1,000, 4,000 and 2,000 calories per unit, respectively and 200, 900 and 500 protein units per unit, respectively. It is required to find the minimum-cost diet containing

at least 20,000 calories and 3,000 units of protein. Formulate and solve the given problem as a LP problem. Write the dual and use it to check the optimal solution of the given problem.

MULTIPLE CHOICE QUESTIONS

1. In standard primal form if the problem is of maximization, all the constraints involve the sign

   (a) $\geq$ (b) $\leq$

   (c) = (d) unrestricted.

2. If the standard primal is of minimization, all the constraints involve the sign

   (a) $\geq$ (b) $\leq$

   (c) = (d) none of these

3. For any primal problem and its dual,

   (a) optimal value of objective functions is same.

   (b) primal will have an optimal solution if and only if dual does too.

   (c) both primal and dual cannot be infeasible.

   (d) all of the above.

4. The right hand side constant of a constraint in a primal problem appears in the corresponding dual as

   (a) a coefficient in the objective function.

   (b) a right hand side constant of a constraint.

   (c) an input-out coefficient.

   (d) none of the above.

5. If a finite optimal feasible solution exists for the primal then the dual has

   (a) unbounded solution

   (b) no solution

   (c) a finite feasible optimal solution

   (d) none of these

6. If the primal problem has an unbounded solution, the dual problem has

   (a) a finite optimal feasible solution

   (b) no solution

(c) either no solution or an unbounded solution

(d) none of these

7. If the ith primal variable of the primal is positive, then the ith variable of the dual is

   (a) +ive (b) –ve

   (c) zero (d) unrestricted

8. If both the primal and dual problems have finite optimal solution and Z_P. Z_D are the optimal values of the objective function of the primal and dual respectively then we have

   (a) $Z_P > Z_D$ (b) $Z_P < Z_D$

   (c) $Z_P = Z_D$ (d) none of these

9. The dual of the primal maximization LP problem having m constraints and n non-negative variables should

   (a) have n constraints and m non-negative variables.

   (b) be a minimization LP problem.

   (c) both (a) and (b).

   (d) none of the above.

10. If at the optimality a primal constraint has positive value of slack variable, then

    (a) dual variable corresponding to that constraint has zero value.

    (b) corresponding resource is not completely used up.

    (c) corresponding resource have zero opportunity cost.

    (d) both (b) and (c) but not (a).

11. The shadow price is

    (a) the price that is paid for purchase of resource.

    (b) the saving by eliminating one of the excess quantities of resource.

    (c) the increase in the objective function value by providing one additional unit of resource.

    (d) none of the above.

12. The value of dual variable

    (a) represents marginal profit of each additional unit of resource.

    (b) can be obtained by examining the z_j row of primal optimal simplex table.

(c) can be obtained by examining $c_j = z_j$ row of primal optimal simplex table.

(d) all of the above.

13. Dual LP problem approach attempts to optimize resource allocation by ensuring that

(a) marginal opportunity cost of a resource equals its marginal return.

(b) marginal opportunity cost of a resource is less than its marginal return.

(c) both (a) and (b).

(d) none of the above.

14. Shadow price indicates how much one unit change in the resource value will change the

(a) optimality range of an objective function.

(b) optimal value of the objective function.

(c) value of the basic variable in the optima) solution.

(d) none of the above.

15. Principle of complementary slackness states that

(a) primal slack x dual main = 0.

(b) primal main x dual surplus = 0.

(c) both (a) and (b).

(d) none of the above.

16. If dual has an unbounded solution, primal has

(a) no feasible solution.

(b) unbounded solution.

(c) feasible solution.

(d) none of the above.

4

Goal Programming

INTRODUCTION

Linear programming technique is applicable only in the situation of single objective (goal) such as minimization of cost/loss or maximization of profit. However, in practice, organizational objectives may vary depending upon the characteristics and philosophy of the organization, statutory regulations, environmental conditions, etc. Though, profit maximization is regarded as the sole objective of the management, in the present dynamic environment of business competition, the organizations have a set of multiple goals. In many situations, the organizations give higher emphasis on goals other than profit maximization. For example, organizations give emphasis on employment stability, social contributions, industrial and labour relations, product quality, revenue maximization, etc.

Goal programming is an approach uised for solving a multi-objective optimization problem that balances trade-offs in conflicting objectives. In other words, it is an approach of deriving a best possible 'satisfactory' level of goal attainment. A problem is modelled into a GP model in a manner similar to that of the LP model. However, the GP model accommodates multiple and often conflicting incommensurable goals in a particular priority order (hierarchy). A particular priority structure is established by ranking and weighing various goals and their subgoals in accordance with their importance. The priority structure helps to deal with all goals (objectives) which cannot be completely and/or simultaneously achieved, in such a manner that more important goals are achieved first at the expense of less important goals.

The management has the situations of multiple conflicting goals and subgoals. This calls for the decision criterion to be multi-dimensional involving multiple goals. In the situation of multiple goals, *Goal Programming* becomes an important quantitative technique, which can handle multiple decisions criteria. These goals are inconsistent in nature and are either in conflict or are achievable only at the cost of other goals. In order to find the solution of the problem, a hierarchy of importance among the inconsistent goals should be established. Hence, low-ordered goals are considered only after the higher ordered goals are satisfied (or further improvement is not desired). If ordinal ranking of goals can be provided in terms of importance or contributions and all goal constraints are linear in nature, the solution of the problem can be obtained through *Goal Programming.*

Optimization of multiple goals requires a different technique for decision-making known as *Goal Programming.* This is a specific technique, developed by Charnes and Cooper, and employs the linear programming for determining the optimum solution of the single or multi-dimensional objective function with a given set of constraints expressed in linear form. In the technique, all the objectives (goals) are incorporated into the objective function and only the conditions, being outside the management's control, are considered as constraints. Further, each goal is set at satisfying level, which may not be the best one obtainable, but management should be satisfied with this to achieve given multiple goals (sometimes may be conflicting).

In goal programming, the deviation between goals and achievables within the given set of constraints are to be minimized. Thus, the objective function contains mainly the deviational variables representing each goal or sub-goal. The deviational variable is represented in two dimensions (a positive and a negative deviation from each sub-goal and/or constraint) in the objective function. As a result, the objective function becomes the minimization of the deviations depending on the relative priority attributed to them.

The solution through linear programming provides the answers to the problems where the variables in cardinal numbers can be quantified accurately. Therefore, the distinguishing feature of goal programming is that it provides for ordinal solution, *i.e.*, where utility or cost of goal or sub-goal may not be specified. However, upper or lower limits for each sub-goal may be specified. Hence, for decision-making, it will be desired to specify the priority and ranking of the required attainment of each goal or sub-goal in ordinal sequence.

Thus, a computational procedure in goal programming is to select a set of solutions satisfying the environmental constraints and giving a satisfactory goal ranked in priority order. In this technique, management sets some estimated targets for each of the goals with the ranking of priorities or importance. Persons desiring to use goal programming only have to specify that which goal is more important than another and not to specify that how much more important. With this information, the goal programming attempts to minimize the deviations from the set target. The computation starts with the most important goal and continues until the achievement of a less priority goal causing the management unsuccessful in achieving a more important one.

DIFFERENCE BETWEEN LP AND GP APPROACH

Linear programming has two major limitations from its applications point of view:

(i) Single objective function, and

(ii) Same unit of measurement of various resources.

The LP model has a single objective function to be optimized such as profit maximization, cost minimization, etc. However, in actual practice, the decision-maker may not strive for a single objective. That is, he may desire to get simultaneous solutions to a complex system of competing objectives.

The solution of any LP problem is based on the cardinal value (the number that expresses exact amount such as, 1, 2, 3, ...) such as profit or cost, whereas GP allows an ordinal solution. Since, it may not be possible to obtain information about the value or cost of a goal (the specific numerical target value desired to achieve) or a subgoal, therefore their upper and lower limits are determined. Usually, priority of a desired goal is assigned and then these priorities are ranked in an ordinal sequence.

Whenever, there are multiple incommensurable (different units of measurement) goals, LP incorporates only one of these goals in the objective function and treats the remaining goals as constraints. Since the optimal solution must satisfy all the constraints, this implies that :

(a) The several goals within the constraining equations are of equal importance, and

(b) These goals have absolute priority over the goal incorporated r into the objective function.

In goal programming, goals are given an ordinal ranking in terms of their contribution or importance to the organization.

FORMULATION OF GOAL PROGRAMMING MODEL

The key to successful application of goal programming to a problem solving is the ability to recognize the possibility of goal programming's application to the problem and formulation of the corresponding model. Following steps are involved in formulation of goal programming model.

1. Developing the objective function.

2. Defining variables and constraints.
3. Formulating constraints.

Developing the Objective Function

The objective function is developed through the analysis of the goal structure. It may be carried out in the following two steps:

(a) The preemptive priority factors should be assigned to specific deviational variables, which are relevant to goal achievement.

(b) If required, differential weights should be attributed to deviational variables at the same level of importance. It is important that goals at the same priority level should be observed.

Defining Variables and Constraints

First of all, the choice variables and the right hand side constraints are defined. The right hand side constraints may be either specified goal targets or available resources. A careful analysis of the problem is required in order to identify all related variables having some influence on the set of goals considered by the decision maker.

Formulating Constraints

A set of constraints should be formulated through an analysis of the relationship among choice variables and their effect on goals. A constraint may be either a goal constraint defining the relationship between choice variables and goals, or a system constraint showing a relationship among the variables. However, if there is no deviational variable to minimize for achievement of a certain goal, a new constraint should be created. Further, a situation of requiring more refinement of goals and priorities can be facilitated by decomposing specific deviational variables. Some examples of formulating constraints are given below:

CONCEPT OF GOAL PROGRAMMING

"The concept of GP was introduced by Channes and Cooper (1961). They suggested a method for solving infeasible LP problems arising from various conflicting resource constraints (goals). A few examples of multiple conflicting goals are:

(i) Maximize profit and increase wages paid to employees,

(ii) Upgrade product quality and reduce product cost,

(iii) Reduce credit losses and increase sales. Ijiri (1965) developed the concept of pre-emptive priority factors, assigning different priority levels to incommensurable goals and different weights for the goals at the same priority level. *Lee* (1972) and *Ignizio* (1976) have written textbooks on the subject of goal programming. Goal programming has been applied to a wide range of planning, resource allocation, policy analysis and functional management problems.

The distinguishing feature of GP is that the goals (targets that the decision-maker would ideally like to achieve for each goal) are satisfied in ordinal sequence. That is, the solution of the GP problem involves achieving some higher goals first, before the lower order goals are considered. Since it is not possible to achieve every goal to the extent desired by the decision-maker, attempts are made to achieve a satisfactory level of all his goals rather than optimum solution for a single goal.

In GP, instead of trying to minimize or maximize the objective function directly as in LP, the deviations from established goals within the given set of constraints are minimized. In the simplex algorithm of linear programming such deviational variables are called slack variables and they are used only as dummy variables. In GP, these slack variables take on a new significance. The deviational variables are represented in two dimensions, both positive and negative deviations from each goal and subgoal. The objective function then becomes the minimization of a sum of these deviations based on the relative importance within the preemptive priority structure assigned to each deviation.

Single Goal with Multiple Subgoals

An objective (goal) is the result desired by a decision-maker. The goal may be underachieved, fully achieved, or overachieved within the given decision environment. The degree of goal achievement depends upon the relative managerial effort applied to an activity. Mathematically, one unit of effort applied to activity x. might contribute an amount a; toward the jth goal.

If the target level for the fth goal is fully achieved, the rth constraint is written as:

$$\sum_{j=1}^{n} a_{ij} x_j = b_i$$

To allow underachievement or overachievement in the target value (goal), let

D_1^- = negative deviation from ith goal (underachievement)

D_i^+ = positive deviation from ith goal (overachievement)

Using these notations, the above stated ith goal can be rewritten as:

$$\sum_{j=1}^{n} a_{ij} x_j + D_i^- - D_i^+ = b_i; \; i = 1, 2, \ldots, m$$

Since both under- and overachievement of a goal cannot be achieved simultaneously, one or both of these deviational variables (D_i^- or D_i^+) be zero in the solution, *i.e.*, $D_i^- \times D_i^+ = 0$. In other words optimality, if one assumes a positive value in the solution, the other must be zero and *vice-versa.*

Remark : The deviational variables in goal programming model are equivalent to slack and surplus variables (the amount by which the objective is below or above the target) in linear programming model.

The deviational variable D_i^+ (called surplus variable in LP) is removed from the objective function of GP when overachievement is acceptable. Similarly, if underachievement is acceptable, D_i^- (called slack variable in LP) is removed from the objective function of GP. But if exact attainment of the goal is desired, then both D_i^- and D_i^+ are included in the objective function and ranked according to their pre-emptive priority factor, from the most important to the least important.

Ranking and Weighting of Unequal Multiple Goals

Since multiple and conflicting goals are usually not of equal rank (importance), deviations (negative and/or positive) from these goals are not additive. Hence to achieve these goals according to their importance a *pre-emptive* priority factor P_1, P_2,... and so on is assigned to goal deviations in the formulation of the objective function to be minimized. The P's do not assume numerical value, they are simply a convenient way of indicating that one goal is more important than another. Priority ranking is absolute and, therefore, these, priority factors have the relationship of $P_1 >> P_2 >> \ldots P_k >> P_{k+1} \ldots$, where >> means 'more important than'. This means, $P_j >> nP_{j+1}$ $(j = 1, 2, \ldots, k)$ where n is a very large number implies that multiplication by n cannot make a lower order goal as important as the higher order goal. Thus, a lowerpriority goal will never be achieved at the expense of a higher priority goal.

It is possible that two or more goals may be assigned equal priority factor (*i.e.*, they are equal in importance) Further, within a given priority there may be subgoals of unequal importance which must be given due importance (weightage). For example, if 'profit' and 'overtime' are assigned same priority factor (*i.e.,* same rank or importance), the decision-maker could assign a weight ranking coefficient, say 2, 4, 5, for overtime duration (in minutes or hours) to reflect worth (measured in Rs) of overtime per minute (or hour) In other words, the differential weights are assigned to the individual deviational variable with the identical priority factor in the GP objective function. It is important to note that the deviational variables at the same priority level must have the same unit of measurement (commensurable), although, deviations that are at different priority levels need not be commensurable.

General GP Model

With m goals, the general goal linear programming model may be stated as:

$$\text{Minimize } Z = \sum_{i=1}^{m}\sum_{r=1}^{k} P_r\left(W_i^- \; D_i^- + W_i^+ \; D_i^k\right)$$

subject to the linear constraints

$$\sum_{j=1}^{m} a_{ij}\, x_j + D_i^- - D_i^+ = b_i; i = 1, 2, ..., m$$

and $x_j, D_i^-, D_i^+ \geq 0$, for all i and j

$$D_i^-, \times D_i^+ = 0,$$

where Z is the sum of the deviations from all desired goals having m goals constraints and n decision variables, x_j. The w_j are non-negative constants representing the relative weight to be assigned to the deviational variables D_i^-, D_i^+ for each goal constraint. The P_r. are the preemptive priorities assigned to the sets of goals that are grouped together in the problem formulation. The a_{ij} are constants attached to each decision variable and the b_i are the right-hand side values (*i.e.,* goals) of each constraint.

FORMULATION OF THE CONSTRAINTS

Sales Goal

Let

x_1 = Total full time salesmen hours per month

x_2 = Total part time salesmen hours per month

D_1^- = Over achievement of sales goal

D_1^+ = Under achievement of sales goal

Given that:

7 = Sales per hour per full-time salesman

3 = Sales per hour per part-time salesman

8000 = Sales goal per month

The sales goal of 8000 cassettes per month is a function of total working hours of the full time and part time salesmen and their output rate (sales/hour). Therefore, the sales goal of 8000 can be represented as following:

$$7x_1 + 3x_2 + D_1^- - D_1^+ = 8000.$$

Regular Working Hours

Let,

D_2^- = Under utilization of the total regular full-time salesmen hours per month.

D_2^+ = Overtime given to full-time salesmen per month.

D_3^- = Under utilization of the total regular part-time salesmen hours per month.

D_3^+ = Extra working hours given to part-time salesmen.

Total regular working hours with the full-time salesmen

$$= 4 \times 150 = 600$$

Total salesmen hour per month with part-time salesmen = 3 × 75 = 225

Thus, the formation of the problem can be given as follows:

$$x_1 + D_2^- - D_2^+ = 600$$
$$x_2 + D_3^- - D_3^+ = 225.$$

Over Time

The second goal is to limit the over-time of full-time salesmen to 80 hours per month. In the above formulated constraints, there is no deviational variable for minimization. Hence, a new constraint is introduced. Now, it is evident that this is the situation of decomposing a certain goal. It may be noted here that the minimization of over-time for full-time salesmen is listed as part of the fourth goal. Therefore, in fact, there are two separate goals regarding the overtime work for full-time salesmen. In order to limit the over-time of full-time salesmen to 80 hours per month, following constraint is introduced:

$$D_2^+ + D_{21}^- - D_{21}^+ = 80$$

D_2^+ = Real overtime of full-time salesman.

D_{21}^- = Difference between the real over-time of full-time salesmen and desired over-time of 80 hours.

D_{21}^+ = Excess over-time over desired 80 hours.

Both the negative and positive deviations from the allowed 80 hours are introduced as the actual over-time can be equal, less than, or even more than 80 hours. Therefore, D_2^+ is the deviation variable to achieve the second goal.

As a result, the goal programming problem can be formulated as follows:

$$\text{Minimize } Z = P_1D_1^- + P_2D_{21}^+ + 2P_3D_2^- + P_3D_3^- + 26.5/11P_4D_3^+ + P_4D_2^+$$

Subject to

$$7x_1 + 3x_2 + D_1^- - D_1^+ = 8000 \text{ (Sales Goal)}$$

$$x_1 + D_2^- - D_2^+ = 600$$

$$x_2 + D_3^- - D_3^+ = 225$$

$$D_2^+ + D_{21}^- - D_{21}^+ = 80$$

Assuming tach full time salesman will charge 50% extra for over-time over the normal wages, *i.e.*, 15 + 7.5 = 22.5

The marginal profit per hour of over-time for full-time salesman

$$= 7 \times 7 - 22.5 = 26.5$$

The marginal profit per hour for part-time salesman = 3 × 7 – 10 = 11.

Thus, the relative cost of an hour of over-time for the part-time salesman is 26.5/11 times that of the full-time salesman. Therefore, (26.5/11) P_4 is assigned to D_3^+ whereas P_4 is assigned to D_2^+.

The productivity ratio (sales per hour) between the full-time and part-time salesmen is 7 to 3, while the hourly wage, rate for over-time is Rs. 22.5 and 10.

GOAL PROGRAMMING THROUGH GRAPHICAL METHOD

The graphical technique for solving goal programming is almost identical to that used for linear programming problems. In linear programming, the graphical technique is used to maximize or minimize an objective function with one goal, whereas in goal programming, the technique is used to minimize the total deviation from a set of multiple goals. The deviation from the goal with highest importance (priority) is minimized to the maximum possible extent before minimizing the deviation from the goal next in priority.

GOAL PROGRAMMING THROUGH SIMPEX METHOD

The simplex method for solving a GP problem is similar to that of an LP problem. The features of the simplex method for the GP problem are:

1. If the target values of each goal in X_B-column is zero, the solution is optimal.
2. To determine the variable to enter into the new solution mix, start examining $(c_j - z_j)$ row of highest priority (P_1) and select the largest negative value. Otherwise, move to the next higher priority (P_2) and select the largest negative value.
3. Apply usual procedure to calculate the 'minimum ratio' to choose a variable to leave the current solution mix (basis).
4. Any negative value in the $(c_j - z_j)$ row which has positive $(c_j - z_j)$ value under any lower priority rows are ignored. It is because deviations from the highest priority goal would be increased with the entry of this variable in the solution mix.
5. The z. and $c_j - z_j$ values are computed separately for each of the ranked goals, P_1, P_2,... . It is because the different goals are measured in different units. These are shown from bottom to top, *i.e.,* first priority goal (P_1) is shown at the bottom and least priority goal at the top.

   The optimality criterion z_j or $c_j - z_j$ becomes a matrix of $k \times n$ size, where k represents the number of pre-emptive priority levels and n is the number of variables including both decision and deviational variables.
6. Examine $c_j - z_j$ values in the P_1-row first. If all $c_j - z_j \le 0$ at the highest priority levels in the same column, then the optimal solution been obtained.

   If $c_j - z_j > 0$ at a certain priority level and there is no negative entry at higher unachieved priority levels, in the same column, the current solution is not optimal.

SOLVED EXAMPLES

Example 1 :

A manufacturing firm produces two types of products: A and B. The unit profit from product A is Rs. 100 and that of product B Rs. 50. The goal of the firm is to earn a total profit of exactly Rs. 700 in the next week.

Solution:

Let us interpret the profit goal in terms of subgoals, which are sales volumes of products: A and B.

x_1 and x_2 = number of units of product A and B produced, respectively

The single goal of profit maximization is stated as:

$$\text{Maximize } Z = 100x_1 + 50x_2$$

Since the goal of the firm is to earn profit with a target established at Rs. 700 per week, the above single goal can be restated to allow for underachievement or overachievement as:

$$100x_1 + 50x_2 + D_1^- - D_1^+ = 700.$$

Now the goal programming model can be formulated as follows:

$$\text{Minimize } Z = D_1^- + D_1^+$$

subject to the constraints

$$100x_1 + 50x_2 + D_1^- - D_1^+ = 700$$

and $$x_1, x_2, D_1^-, D_1^+ \geq 0$$

where D_1^- = underachievement of the profit goal of Rs. 700

D_1^+ = overachievement of the profit goal of Rs. 700.

If the profit goal is not completely achieved, the slack in the profit goal will be expressed by negative deviational (under-achievement) variable, df from the goal. But, if the solution shows a profit in excess of Rs. 700, the surplus in the profit will be expressed by positive deviational (overachievement) variable D_1^+ from the goal. If the profit goal of Rs. 700 is exactly achieved, both D_1^- and D_1^+ will be zero.

There are an infinite number of combinations of x_1 and x_2 that will achieve the profit goal. The required solution will be any linear combination of x_1 and x_2 between the two points: $x_1 = 7, x_2 = 0$ and $x_1 = 0, x_2 = 14$. This straight line is exactly the iso-profit function line when total profit is Rs 700.

Example 2:

In Example 1, let us suppose that the manager in addition to the profit goal of Rs. 700 also wants to achieve sales volume for product A and B close to 5 and 4, respectively. Formulate this problem as a goal programming model.

Solution:

The constraints of the problem can be stated as:

$$100x_1 + 50x_2 = 700 \text{ (profit target goal)}$$

$$x_1 \leq 5$$

$$x_1 \leq 4\} \text{ (sales target goals)}$$

The problem can now be formulated as GP model as follows:

Minimize $Z = D_1^- + D_1^+ + D_2^- + D_3^-$

subject to the constraints

$$100x_1 + 50x_2 + D_1^- - D_1^+ = 700$$

$$x_1 + D_2^- = 5$$

$$x_2 + D_3^- = 4$$

and $\quad x_1, x_2, D_1^-, D_1^+, D_2^- + D_3^- \geq 0$

where D_2^- and D_3^- represent underachievements of sales volume for product A and B, respectively. Since sales target goals are given as the maximum possible sales volume, therefore, D_2^+ and D_3^+ are not included in the sales target constraints.

The solution to this problem can be found by a simple examination of the problem. If $x_1 = 5$ and $x_2 = 4$, then all targets will be completely achieved. Thus $D_1^- = + D_2^- = D_3^- = D_1^+ = 0$.

Example 3:

A manager of the only audio cassettes shop has a decision problem which involves multiple goals. The audio cassettes shop employs 4 full time and 3 part time salesmen having normal working hours per month 150 and 75, respectively. According to the performance records, the average sale has been 7 cassettes per hour for full time salesman and 3 cassettes per hour for part time salesman. The average hourly wage rates are Rs. 15 for full time salesman and Rs. 10 for part time salesman. Average profit from the sales of a cassette is Rs. 7. In view of various considerations, the manager sets the following goals in order of priority.

1. *To achieve a sales goal of 8000 cassettes per month.*
2. *To limit the over-time of full time salesmen to 80 hours per month.*
3. *To provide job security to salesmen by full utilization of employee's regular working hours, but concerned twice with the full utilization of full time salesmen as with the part time salesmen.*

4. To minimize the sum of over-time for both categories of salesmen with the desire to assign differential weights to the minimization of over-time based on net marginal profit ratio between the full time and part-time salesmen.

Solution:

Do yourself as above example.

Example 4:

A company manufactures two products by using two machines in sequence. The time available for the two products on each machine is limited to 10 hours daily but can be exceeded by 5 hours on an over time basis. Each overtime hour costs an additional Rs. 50. The production rates with the profit for the two products are given as follows:

Production Rate

Machine	*Product-1*	*Product-2*
1	*8*	*10*
2	*5*	*13*
Profit per unit (Rs.)	*40*	*60*

Formulate the goal programming model for determining the production level for each product for maximum profit.

Solution:

Variables are number of units of each product which are to be determined

Objective is maximization of net profit

Constraints are maximum allowable hours with over-time hours.

If

x_1 = number of units of product-1.

x_2 = number of units of product-2.

In the absence of over-time option, the constraints of the model can be written as

$$\frac{x_1}{8} + \frac{x_2}{10} \leq 10 \text{ (machine-1)}$$

$$\frac{x_1}{5} + \frac{x_2}{13} \leq 10 \text{ (machine-2)}$$

To include the overtime option, the constraints can be written as

$$\frac{x_1}{8} + \frac{x_2}{10} - y_1 = 10$$

$$\frac{x_1}{5} + \frac{x_2}{13} - y_2 = 10.$$

Where y_1 and y_2 are the overtime hours on machine-1 and machine-2, respectively.

y_1 and y_2 are unrestricted in sign for the following reasons:

(a) If y_i $(i = 1, 2)$ is negative, the 10 hour limit is not exceeded and no over-time is used.

(b) If y_i $(i = 1, 2)$ is positive, the used machine hours will exceed the daily limit and y_i will represent the over-time hours.

Next step is to limit the daily use of over-time to 5 hours and to include the over-time cost in the objective function.

The constraint $y_i \leq 5$ (when overtime is used) will provide the desired restriction on the use of over-time. The constraint become redudant when $y_i < 0$ (no over-time).

The objective is to maximize the net profit which is equal to total profit from the two products minus the additional cost of over-time. The total profit is given by

$$40x_1 + 60x_2.$$

The over-time cost can be expressed as,

Over-time cost = Cost per hour × over-time hours = 50 {max. $(0, y_i)$}

It may be noted that max. $(0, y_i) = 0$ when $y_i < 0$, *i.e.*, yielding zero over-time cost as desired.

The complete goal programming model for Example 1 can be given as following:

Maximize $40x_1 + 60x_2 - 50\ [\max\ (0, y_1) + (0, y_2)]$

Subject to

$$\frac{x_1}{8} + \frac{x_2}{10} - y_1 = 10$$

$$\frac{x_1}{5} + \frac{x_2}{13} - y_2 = 10.$$

$$x_1, x_2 \geq 0$$

y_1, y_2 unrestricted in sign

The model can be converted to linear program by substituting w_i = max $(0, y_i)$ which is equivalent to

$$w_i \geq y_i \text{ and } w_i \geq 0$$

Since the coefficient of w_i in objective function is negative, its value should be assumed smallest possible non-negative, *i.e.*, zero or y_i. Thus, the Linear Program Model can be given as following:

Maximize $Z = 40x_1 + 60x_2 - 50(w_1 + w_2)$

Subject to

$$\begin{aligned}
x_1/8 + x_2/10 - y_1 &= 10 \\
x_1/5 + x_2/13 - y_2 &= 10 \\
y_1 - w_1 &\leq 0 \\
y_2 - w_2 &\leq 0 \\
y_1 &\leq 5 \\
y_2 &\leq 5 \\
x_1, -x_2, w_1, w_2 &\geq 0
\end{aligned}$$

y_1, y_2, unrestricted in sign.

Example 5:

An office equipment manufacturer produces two kinds of products, chairs and lamps. Production of either a chair or a lamp requires one hour of production capacity in the plant. The plant has a maximum production capacity of 50 hours per week. Because of the limited sales capacity, the maximum number of chairs and lamps that can be sold are 6 and 8 per week, respectively. The gross margin from the sale of a chair is Rs. 90 and Rs. 60 for a lamp.

The plant manager desires to determine the number of units of each product that should be produced per week in consideration of the following set of goals:

Goal 1: Available production capacity should be utilized as much as possible but not exceeded.

Goal 2: Sales of two products should be as much as possible.

Goal 3: Overtime should not exceed 20 per cent of available production time.

Formulate and solve this problem as a GP model so that the plant manager may achieve his goals as closely as possible.

Solution:

Let x_1 and x_2 = number of units of chair and lamp produced, respectively.

The first goal pertains to production capacity attainment with a target established at 50 hours per week. This constraint can be stated as:

$$x_1 + x_2 + D_1^- - D_1^+ = 50$$

where D_1^- = underutilization (idle) of production capacity

D_1^+ = overutilization (overtime) of production capacity.

If this goal is not achieved, D_1^- would take on a positive value and D_1^+ would be zero.

The second goal pertains to maximization of sales which have sales volume constraints. The sales constraints can be expressed as:

$$x_1 + D_2^- = 6$$

$$x_1 + D_3^- = 8$$

Since the sales goals are the maximum possible sales volume, D_2^+ and D_3^+ will not appear in these constraints. Thus, overachievement of sales goals is ruled out.

The third goal pertains to the minimization of overtime as much as possible. The constraint is stated as follows:

$$D_1^+ + D_4^- - D_4^+ = 0.2\ (50) = 10$$

where D_4^- = overtime less than 20 per cent of goal constraint

D_4^+ = overtime more than 20 per cent of goal constraint

D_1^+ = overtime beyond 50 hours.

Now, the given problem can be stated as a goal programming model as:

Minimize (total deviation) $Z = D_1^+ + D_2^- + D_3^- + D_4^+$

subject to the constraints

(i) Production capacity constraint

$$x_1 + x_2 + D_1^- - D_1^+ = 50$$

(ii) Sales constraints

$$x_1 + D_2^- = 6$$

$$x_2 + D_3^- = 8$$

(iii) Overtime constraint

$$D_1^+ + D_4^- - D_4^+ = 10$$

and $x_1, x_2, D_1^-, D_1^+, D_2^-, D_3^-, D_4^-, D_4^+, \geq 0.$

Example 6:

A manufacturing firm produces two products A and B. Each unit of the product-A requires 4 hours of labour and 6 units of raw material, whereas each unit of product-B requires 8 hours of labour and 8 units of raw material. Each day 40 hours of labour and 48 units of raw material are available. The profit per unit of products-A and B are Rs. 160 and Rs. 120, respectively.

The goal priorities of management are as following:

1. *The profit per day should be atleast Rs. 720.*
2. *The number of units of product-A should be twice the number of units of product-B.*
3. *Full utilization of labour should be made.*

Solution

Let

x_1 = Quantity of product-A.

x_2 = Quantity of product-B.

Then the constraints of the problem can be written as follows:

$$160x_1 + 120x_2 \geq 720 \text{ (profit)}$$
$$x_1 - 2x_2 = 0 \text{ (production balance)}$$
$$4x_1 + 8x_2 \leq 40 \text{ (labour constraint)} \qquad \ldots (1)$$
$$6x_1 + 8x_2 \leq 48 \text{ (raw material constraint)}$$

Now introducing two non-negative constraints to the first two constraints of (1) we get

$$160x_1 + 120x_2 - D_1^- + D_1^+ = 720$$
$$x_1 - 2x_2 - D_2^- + \quad = 0 \qquad \ldots (2)$$

Where D_1^+ = Over achievement of *i*th goal

and D_1^- = Under achievement of *i*th goal

$i = 1, 2$

Since for 3rd and 4th constraints of (1), over-achievement is zero (by definition), only D_3^- and D_4^- are used as D_3^+ and D_4^+ are zero.

$$4x_1 + 8x_2 + D_3^- = 40$$

$$6x_1 + 8x_2 + D_4^- = 48 \quad \ldots (3)$$

Now the complete model can be written as,

$$160x_1 + 120x_2 - D_1^+ + D_1^- = 720$$

$$x_1 + 2x_2 - D_2^+ + D_2^- = 0$$

$$4x_1 + 8x_2 + D_3^- = 40$$

$$6x_1 + 8x_2 + D_4^- = 48$$

$$x_1, x_2, D_1^-, D_1^+, D_2^-, D_2^+, D_3^-, D_3^+, D_4^- \geq 0.$$

The next step is to analyze each goal in the objective function in finding the optimal solution in the feasible region. The first goal is completely achieved at line PE in the feasible region PSDE.

Now, moving to priority level-2 which is completely achieved at line OM in the shaded region. Now, moving for priority-3, differential weight for product-A is 2 times that of product-B. The line AB representing the labour utilization and line OM representing production balance interact outside the feasible region. The third goal (full utilization of labour) cannot be achieved without dispensing with first two goals in the priority order. The optimal solution can be given at point "R", *i.e.*,

$$x_1 = \frac{24}{5}, x_2 = \frac{12}{5}$$

The total profit $= 160 \times \frac{24}{5} + 120 \times \frac{12}{5} = 1056$

and thus goal-1 is over achieved with the quantum of over-achievement

$$D_1^+ = 336$$

Goal-2 has been achieved, since $D_2^+ = D_2^- = 0$

and $\quad x_1 = 2x_2$

Goal-3 has not been achieved. The degree of under achievement

$$D_3^- = 8/5 \text{ units.}$$

Thus, the above solution is the best attainable under existing constraints.

Example 7:

Entertainment System Ltd. manufactures two products, TVs and Stereo sets, which are processed through assembly and finishing sections. Assembly has 60 hours available per week, whereas finishing can handle upto 48 hours a week. Manufacturing of 1 TV set requires 4 hours in assembly and 2 hours in finishing. Each stereo set requires 2 hours in assembly and 4 hours in finishing. The profit is 8 units (Rs. 100) per TV set and 6 units (Rs. 100) per stereo set. Under the situation of equally ranked goals, the firm states following two goals:

1. *To reach a profit target of 100 units*
2. *To meet a TV set goal of 10*

Since these goals are equally ranked, a unit deviation from the profit target is just as important as a deviation of 1 TV set.

Under the situation of goals ranked in priority, the firm states the following:

1. *Produce to meet a TVs goal of 13 (P_1).*
2. *Reach a profit goal of 135 units (P_2).*
3. *Produce to meet a stereos goal of 5 (P_3).*

Compute the best solution through goal programming technique under the above two situations.

Solution

A. Situation of two equally ranked goals

Let

x_1 and x_2 are the quantities of TVs and Stereos respectively and

D_1^- = Amount by which the profit goal is underachieved.

D_1^+ = Amount by which the profit goal is overachieved.

D_2^- = Amount by which the TV sets goal is underachieved.

D_2^+ = Amount by which the TV sets goal is overachieved.

D_3^- = Amount by which the stereo sets goal is underachieved.

D_3^+ = Amount by which the stereo sets goal is overachieved.

The objective function and constraints for the solution to this problem can be given as,

Minimize $D_1^- + D_1^+$

Subject to

$$8x_1 + 6x_2 + D_1^- + D_1^+ = 100 \quad \text{(Profit goal)}$$

$$x_1 + D_2^- - D_2^+ = 10 \qquad \text{(TVs goal)}$$

$$4x_1 + 2x_2 \leq 60 \qquad \text{(Assembly constraint)}$$

$$2x_1 + 4x_2 \leq 48 \qquad \text{(Finishing constraint)}$$

$$x_1,\ x_2,\ D_1^-,\ D_1^+,\ D_2^-,\ D_2^+ \geq 0$$

The slack variables S_1 and S_2 can be introduced in constraint equations, *i.e.*,

$$4x_1 + 2x_2 + S_1 = 60$$

$$2x_1 + 4x_2 + S_2 = 48.$$

The initial solution and optimum solution to this goal programming problem are shown in Table 4.1. From the Table 7.1, it can be observed that the goal of 10 TVs is overachieved by $2\frac{1}{2}\left(12\frac{1}{2} - 10\right)$. $2\frac{1}{2}$ appears in the final solution as D_2^+ (overachievement of TVs). The profit goal of 100 units was reached (both D_1^- and D_1^+ are zero as they are not in final solution and, thus, the profit is exactly 100 units).

Table 4.1 : Initial Final Solution to Example 4.4 with two Equally Ranked Goals

C_j	*Product Mix*	*Quality*		*0* x_1	*0* x_2	*0* S_1	*0* S_2	*1* D^-_1	*0* D^+_1	*1* D^-_2	*0* D^+_2
1	D^-_1	100		8	6	0	0	1	–1	0	0
1	D^-_2	10	Initial	1	0	0	0	0	0	1	–1
0	S_1	60	Solution	4	2	1	0	0	0	0	0
0	S_2	48		2	4	0	1	0	0	0	0
	Z_j	110		9	–6	0	0	1	–1	1	–1
	$C_j - Z_j$			–9	–6	0	0	0	1	0	1
0	D^+_2	2½		0	3/4	0	0	1/8	–1/8	–1	1
0	x_1	12½	Optimal	1	3/4	0	0	1/8	–1/8	0	0
0	S_1	10	Solution	0	–1	1	0	–1/2	1/2	0	0
0	S_2	23		2	2½	0	1	–1/4	1/4	0	0
	Z_j			0	0	0	0	0	0	0	0
	$C_j - Z_j$			0	0	0	0	1	0	1	0

Situation of Priority Ranked Multiple Goals

Goal programming can also handle the situation of priority ranked multiple goals. In general, one goal is more important than another.

The solution of problem in such a situation can be illustrated through the same example as considered for equally ranked goals.

Suppose Entertainment Systems Ltd. has established the following goals with priority P_1, P_2, and P_3 where P_1 is most important as follows:

1. Produce to meet a TVs goal of 13 (P_1).
2. Reach a profit goal of 135 units (P_2).
3. Produce to meet a stereos goal of 5 (P_3).

Here, P_1, P_2, and P_3 used are called preemptive priority factors which follows that the priority ranking is absolute, *i.e.*, P_1 goal is so much important than P_2 goal that P_2 goal will never be achieved before the achievement of P_1 goal.

The objective function and constraints for this situation of goal programming problem are

Minimize $P_1\ D_2^- + P_2\ D_1^- + P_3\ D_3^-$

Subject to

$8x_1 + 6x_2 + D_1^- - D_1^+ = 135$ (Profit goal)

$x_1 + D_2^- - D_2^+ = 13$ (TVs goal)

$x_2 + D_3^- - D_3^+ = 5$ (Stereos goal)

$4x_1 + 2x_2 \leq 60$ (Assembly constraint)

$2x_1 + 4x_2 \leq 48$ (Finishing constraint)

$x_1, x_2, D_1^-, D_1^+, D_2^-,$
$D_2^+, D_3^-, D_3^+ \geq 0.$

The constraint equation can be converted to equality equations by introducing slack variables S_1 and S_2 respectively. The equations formed will be as following:

$$4x_1 + 2x_2 + S_1 = 60$$
$$2x_1 + 4x_2 + S_2 = 48.$$

Table 4.2 Presents the initial Simplex Table for the above problem.

Table 4.2 : Initial Simplex Table for Priority Ranked Entertainment Systems Ltd. Problem

C_j	Product Mix	Quantity	0 x_1	0 x_2	0 S_1	0 S_2	P_2 D^-_1	0 D^+_1	P_1 D^-_2	0 D^+_2	P_3 D^-_3	0 D^+_2
P_2	D^-_1	135	8	6	0	0	1	–1	0	0	0	0
P_1	D^-_2	13	1	0	0	0	0	0	1	–1	0	0
P_3	D^-_3	5	0	1	0	0	0	0	0	0	1	–1
0	S_1	60	4	2	1	0	0	0	0	0	0	0
	S_2	48	2	4	0	1	0	0	0	0	0	0
			0	1	0	0	0	0	0	0	1	–1
			0	–1	0	0	0	0	0	0	0	1
			8	0	0	0	1	–1	0	0	0	0
			–8		0	0	0	1	0	0	0	0
			1	0	0	0	0	0	1	–1	0	0
			(–1)	0	0	0	0	0	0	1	0	0
			— In									

Table 4.2 possesses a number of characteristics which are explained below:

1. There is a separate Z_j and $C_j - Z_j$ row for each of the P_1, P_2, and P_3 priorities, Since the deviation from the product goals are not added to those from the profit goal (the priorities are different), separate priority rows are needed to keep the track of situations. Further, it is a common practice to represent priority rows from bottom to top in order of priority.
2. The $C_j - Z_j$ value for any column is given in the priority rows at the bottom of the Table, *i.e.,* the $C_j - Z_j$ value for the x_1 column is contained in the P_2 and P_1 rows at the bottom and is read $-8P_2 - 1P_1$, and, similarly, the $C_j - Z_j$ value for x_2 column is read $-1P_3 - 6P_2$.
3. For selecting the variable to enter the mix (x_1 in this case), it is started with the most important priority P_1 and the most negative $C_j - Z_j$ in that row is picked up. If there is no negative $C_j - Z_j$ in that row, the next most important priority P_2 is considered and looked in that row.
4. Replaced variable is determined by the use of same procedure as used earlier. In this case, 13/1 is the smallest positive value and, thus, the D_2^- row will be replaced in the next Tableau.
5. If the value of $C_j - Z_j$ is *positive* in one of the P-rows underneath it, It is disregarded. Such a positive value would mean that deviations

from the lower (and more important) goal would be increased if that variable is brought in. This is avoided as it will not lead to a better solution.

Table 4.3 : Initial Solution for Priority Ranked Entertainment Systems Ltd. Problem

C_j	*Product Mix*	*Quantity*	0 x_1	0 x_2	0 S_1	0 S_2	P_2 D^-_1	0 D^+_1	P_1 D^-_2	0 D^+_2	P_3 D^-_3	0 D^+_3
P_2	D_1^-	7	0	0	–3	0	1	–1	4	–4	0	0
0	x_1	13	1	0	0	0	0	0	1	–1	0	0
P_3	D_3^-	1	0	0	–1/2	0	0	0	2	–2	1	–1
0	x_2	4	0	1	1/2	0	0	0	–2	2	0	0
	S_2	6	0	0	–2	1	0	0	6	–6	0	0
			0	0	–1/2	0	0	0	2	–2	1	–1
			0	0	1/2	0	0	0	(–2)	2	0	1
			0	0	–3	0	1	–1	4	4	0	0
			0	0	3	0	0	1	–4	4	0	0
			0	0	0	0	0	0	0	0	0	0
			0	0	0	0	0	0	1	0	0	0

Once the initial simplex tableau is formed, the solution procedure is followed as in simplex method, keeping in mind the above five characteristics. The final simplex Tableau for this problem is shown in Table 4.3. It may be noted that $C_j - Z_j$ values would be negative in both the P_2 and P_3 rows (–4 and –2, respectively). However, in both cases, this is a positive value in the P_1 row beneath them (1), and thus both of them are disregarded.

It may be noted from Table 4.3 that the goals have been reached in different extents. The most important goal (produce 13 TVs) has been achieved. The second most important goal (read 135 units of profit) has been missed (only a profit of 128 units is generated). The deviation, however, is slight, *i.e.*, about 5%. This deviation (7) is shown as D_1^- in the final solution (Table 4.3). The lower ranked goal (produce 5 stereos) has been missed by 1 unit which represents a 20% deviation which is shown as D_3^- with a value of 1. It may be observed that there are 6 hours unused is finishing, whereas it may look that these should have been used. However, manufacturing of another stereo would have required time in assembly, where all the time has been used.

EXERCISES

1. ABC Furnitures produce three products, tables, desks and chairs. The furniture is produced in the central plant. The production of the desk requires 3 hours in the plant, the table 2 hours and the chair only one hour. The regular plant capacity is 40 hours a week. According to the marketing department, the maximum number of desks, tables and chairs that can be sold weekly are 10, 10 and 12, respectively. The president of the firm has established the following goals according to their importance:

   (i) Avoid any underutilization of production capacity.

   (ii) Meet the order of XYZ Store for seven desks and five chairs.

   (iii) Avoid overtime operation of the plant beyond 10 hours.

   (iv) Achieve sales goal of 10 desks, 10 tables, and 12 chairs.

   (v) Minimize overtime operation as much as possible.

   Formulate and solve this problem as a goal programming problem.

2. ABC Computer Company produces three different types of computers—Epic, Galaxie and Utopia. The production of all computers is conducted in a complex and modem assembly line. The production of an Epic requires 5 hours in the assembly line, a Galaxie requires 8 hours and a Utopia requires 12 hours. The normal operation hours of assembly line are 170 per month. The marketing and accounting departments have estimated that profits per unit for the three types of computers are Rs. 1,00,000 for the Epic, Rs. 1,44,000 for the Galaxie and Rs. 2,52,000 for the Utopia. The marketing department further reports that the demand is such that the firm can expect to sell all the computers it produces in the month. The chairman of the company has established the following goals listed according to their importance:

   (i) Avoid underutilization of capacity in terms of regular hours of operation of the assembly line.

   (ii) Meet the demand of the north-eastern sales district for five Epics, five Galaxies, and eight Utopias (differential weights should be assigned according to the profit ratios among the three types of computers).

(iii) Limit overtime operation of the assembly line to 20 hours.

(iv) Meet the sales goal for each type of computer: Epic-10; Galaxie-12; and Utopia-10 (again assign weights according to the relative profit function for each computer).

(v) Minimize the total overtime operation of the assembly line.

Formulate and solve this problem as a goal programming model through two iterations (three tables) by the simplexmethod.

3. What is goal programming? Why are all goal programming problems minimization problems? Why does altering the goal priorities result in a different solution to a problem? Explain.
4. What is meant by 'satisfying' and why is the term often used in conjunction with goal programming?
5. What are deviational variables? How do they differ from decision variables in traditional linear programming problems?
6. What does it mean to rank goals in goal programming? How does this affect the problem's solution?
7. What is goal programming? Clearly state its assumptions.
8. (a) Compare the differences/similarities between linear programming and goal programming.

   (b) Why are all goal linear programming problems minimization problems? Explain.
9. Explain the differences in solving a linear programming versus goal programming problem by the simplex method. State some problem areas in management where goal programming might be applicable.
10. 'Goal programming appears to be the most appropriate, flexible and powerful technique for complex decision problems involving multiple conflicting objectives.' Discuss.
11. An office equipment manufacturer produces two kinds of products, chairs and lamps. Production of either a chair or a lamp requires one hour of production capacity in the plant. The plant has a maximum production capacity of 10 hours per week. Because of the limited sales capacity, the maximum number of chairs and lamps that can be sold are 6 and 8 per week, respectively. The gross margin from the sale of a chair is Rs. 80 and Rs. 40 for that of a lamp.

    The plant manager has set the following goals arranged in order of importance:

(i) He wants to avoid any underutilization of production capacity.

(ii) He wants to sell as many chairs and lamps as possible. Since the gross margin from the sale of a chair is set at twice the amount of profit from a lamp, he has twice as much desire to achieve the sales goal for chairs as for lamps.

(iii) He wants to minimize overtime operation of the plant as much as possible.

Formulate and solve this problem as a GP problem, so that the plant manager makes a decision that will achieve his goals as closely as possible.

12. A production manager faces the problem of job allocation among three of his teams. The processing rates of the three teams are 5, 6 and 8 units per hour, respectively. The normal working hours for each team is 8 hours per day. The production manager has the following goals for the next day in order of priority:

(i) The manager wants to avoid any underachievement of production level, which is set at 180 units of product.

(ii) Any overtime operation of team 2 beyond two hours and team 3 beyond three hours should be avoided.

(iii) Minimize the sum of overtime.

Formulate and solve this problem as a goal programming problem.

13. XYZ company produces two products—record players and tape-recorders. The production of both products is done in two separate machine centres within the plant. Each record player requires two hours in machine centre A and one hour in machine centre B. Each tape-recorder, on the other hand, requires one hour in the machine centre A and three hours in machine centre B. In addition, each product requires some in-process inventory. The per-unit in-process inventory required is Rs. 50 for the record player and Rs. 30 for the tape-recorder. The firm has normal monthly operation hours of 120 for both machine centres A and B. The estimated profit per unit is Rs. 100 for the record player and Rs. 75 for the tape-recorder. According to the marketing department, the forecast of sales for the record player and the tape-recorder are 50 and 80, respectively for the coming month.

The president of the firm has established the following goals for production in the next month, in ordinal rank of importance:

(i) Limit the amount tied up in in-process inventory for the month to Rs. 4,600.

(ii) Achieve the sales goal of 80 tape-recorders for the month.

(iii) Limit the overtime operation of machine centre A to 20 hours.

(iv) Achieve the sales goal of 50 record players for the month.

(v) Limit the sum of overtime operation for both machine centres.

(vi) Avoid any underutilization of regular operation hours of both machine centres.

Formulate and solve this problem as a goal programming problem.

14. Explain the following terms

(i) Deviational variables

(ii) Pre-emptive priority factors

15. Explain the difference between cardinal value and ordinal value.

16. Under what circumstances can cardinal weights be used in the objective function of a goal programming model?

17. What happens if cardinal weights are attached to all priorities in the objective function of a goal programming model?

18. The manager of the only record shop in a town has a decision problem that involves multiple goals. The record shop employs five full-time and four part-time salesmen. The normal working hours per month for a full-time salesman are 160 hours and for a part-time salesman 80 hours. According to performance record of the salesmen, the average sales has been five records per hour for full-time salesmen and two records per hour for part-time salesmen. The average hourly wage rates are Rs. 3 for full-time salesmen and Rs. 2 for part-time salesmen. Average profit from the sales of a record is Rs. 1.50. In view of past record of sales, the manager feels that the sales goal for the next month should be 5,500 records. Since the shop is open six days a week, overtime is often required of salesmen (not necessarily overtime but extra hours for the part-time salesmen). The manager believes that a good employer-employee relationship is an essential factor of business success. Therefore, he feels that a stable employment level with occasional overtime requirement is a better practice than an unstable employment level with no overtime. However, he feels that overtime of more than 100 hours among the full-time salesmen should be avoided because of the declining sales effectiveness caused by fatigue. The manager has set the following goals:

(i) The first goal is to achieve a sales target of 5,500 records for the next month.

(ii) The second goal is to limit the overtime of full-time salesmen to 100 hours.

(iii) The third goal is to provide job security to salesmen. The manager feels that full utilization of employees' regular working hours (no layoffs) is an important factor for a good employer-employee relationship. However, he is twice as concerned with the full utilization of full-time salesmen as with the full utilization of part-time salesmen.

(iv) The last goal is to minimize the sum of overtime for both full-time and part-time salesmen. The manager desires to assign differential weights to the minimization of overtime according to the net marginal profit ratio between the full-time and part-time salesmen. Formulate and solve the given problem as a goal programming problem.

19. A hospital administration is reviewing departmental requests prior to the design of a new emergency room. At issue is the number of beds for each department. The current plans call for a 15,000 square feet facility. The hospital board has established the following goals in order of importance:

Department	*No. of Beds Requested*	*Cost per Sed (incl. equipment) (Rs.)*	*Area per bed (sqft)*	*Peak Requirement (Max. no. of patients at one time)*
A	5	12,600	474	3
B	20	5,400	542	18
C	20	8,600	438	15

(i) Avoid overspending of the budget Rs. 3,00,000.

(ii) Avoid plan requiring more than 15,000 sq. ft.

(iii) Meet the peak requirement.

(iv) Meet the departmental requirements.

Formulate and solve (up to three iterations) the given problem as a goal programming problem.

20. Mr. X has inherited Rs. 10,00,000 and seeks your advice concerning investing his money. You have determined that 10 per cent can be earned on a bank account and 14 per cent by investing in certificates

of deposit. In real estate, you estimate an annual return of 15 per cent, while on the stock market you estimate that 20 per cent can be earned annually. Mr. X has established the following goals in order of importance:

(i) Minimize risk; therefore he wants to invest not more than 35 per cent in any one type of investment.

(ii) Must have Rs. 1,00,000 in the bank account to meet any emergency.

(iii) Maximum annual cash return.

Formulate and solve (up to three iterations only) this problem as a goal programming model that determines the amount of money which X should invest in each investment option.

21. A company plans to schedule its annual advertising campaign. The total advertising budget is set at Rs. 10,00,000. The firm can purchase local radio spots at Rs. 2,000 per spot, local television spots at Rs. 12,000 per spot or magazine advertising at Rs. 4,000 per insertion. The payoff from each advertising medium is a function of its audience size and audience characteristics. Let this payoff be defined as audience points. Audience points for the three advertising vehicles are:

Radio	50 points per spot
Television	250 points per spot
Magazine	200 points per insertion.

The Advertising Manager of the firm has established the following goals for the advertising campaign, listed in the order of importance:

(i) The total budget should not exceed Rs. 10,00,000,

(ii) The contract with the radio and television station requires that the firm spend at least Rs. 3,00,000 for television and radio ads.

(iii) The company does not wish to spend more than Rs. 2,00,000 for magazine ads.

(iv) Audience points from the advertising campaign should be maximized.

Formulate and solve (up to three iterations only) this problem as a goal programming problem by the simplex method.

22. A shoe manufacturer produces hiking boots and ski boots. Its manufacturing process consists of sewing and stitching. It has available 60 hours per week for the sewing process and 80 hours per week for the stitching process at normal capacity. The firm realizes profits of

Rs. 150 per pair on hiking boots and Rs. 100 per pair on ski boots. It requires 2 hours of sewing and 5 hours of stitching to produce one pair of hiking boots and 3 hours of sewing and 2 hours of stitching to produce one pair of ski boots. The president of the company wishes to achieve the following goals, listed in the order of their importance:

(i) Achieve the profit goal of Rs. 5,250 per week.

(ii) Limit the overtime operation of the sewing center to 30 hours.

(iii) Meet the sales goal for each type of boot—25 hiking boots and 20 ski boots.

(iv) Avoid any under-utilization of the regular operation hours of the sewing center.

Formulate and solve this problem as a goal programming model.

23. The last goal is to minimize the sum of overtime for both full-time and part-time salesmen. The manager desires to assign differential weights to the minimization of overtime according to the net marginal profit ratio between the full-time and part-time salesmen. Formulate and solve the given problem as a goal programming problem.

24. For selecting the variable to enter the mix (x_1 in this case), it is started with the most important priority P_1 and the most negative $C_j - Z_j$ in that row is picked up. If there is no negative $C_j - Z_j$ in that row, the next most important priority P_2 is considered and looked in that row.

MULTIPLE CHOICE QUESTIONS

1. In Goal Programming problem, goals are assigned priorities such that
   (a) higher priority goals must be achieved before lower priority goals
   (b) goals may not have equal priority
   (c) goals of greatest importance are given lowest priority
   (d) all of the above

2. In GP problem, a constraint having unachieved vari- able is expressed as:
   (a) an equality constraint
   (b) a less than or equal to type constraint
   (c) a greater than or equal to type constraint
   (d) all of the above

3. For applying a Goal Programming approach, decision-maker must
   (a) set targets for each of the goals

(b) assign pre-emptive priority to each goal

(c) assume that linearity exists in the use of resources to achieve goals

(d) all of the above

4. Goal programming

(a) requires only that decision-maker knows whether the goal is direct profit maximization or cost minimization

(b) allows you to have multiple goals, with or without priorities

(c) is an approach to achieve goal of a solution to all IP problems

(d) none of the above

5. In simplex method of goal programming, the variable to enter the solution mix is selected with

(a) lowest priority row and most negative $c_j - z_j$ value in it

(b) lowest priority row and largest positive $c_j - z_j$ value in it

(c) highest priority row and most negative $c_j - z_j$. value in it

(d) highest priority row and most positive $c_j - z_j$. value in it.

6. Consider a goal with constraint: $g_1 (x_1, x_2, ..., x_n) + D_1^- - D_1^+ = b_1$ and the term $3D_1^- + D_1^+$ in the objective function, the decision-maker

(a) prefers $g_1 (x_1, x_2 ..., x_n) \geq b_1$ rather than $\leq b_1$

(b) prefers $g_1 (x_1, x_2, ..., x_n) \leq b_1$ rather than $\geq b_1$

(c) not concerned with either $\leq$ or $\geq$.

(d) none of the above

7. Consider a goal with constraint: $g_1 (x_1, x_2,..., x_n) + D_1^- \geq b_1$ $(D_1^- \geq 0)$ with D_1^- in the objective function. Then

(a) the goal is to minimize underachievement

(b) the constraint is active provided $D_1^- > 0$

(c) both (a) and (b)

(d) none of the above

8. The use of GP model is preferred when

(a) goals are satisfied in an ordinal sequence

(b) goals are multiple incommensurable

(c) more than one objective is set to achieve

(d) all of the above

9. Deviational variables in GP model must satisfy following conditions
   (a) $D_i^+ \times D_i^+ = 0$
   (b) $D_i^+ - D_i^- = 0$
   (c) $D_i^+ + D_i^- = 0$
   (d) none of the above
10. In GP at optimality, which of the following conditions indicated that a goal has been exactly satisfied :
    (a) positive deviational variable is in the solution mix with a negative value
    (b) both positive and negative deviational variables are in the solution mix
    (c) both positive and negative deviational variables are not in the solution mix
    (d) none of the above.

Sequencing Problem

INTRODUCTION

The sequencing problem arises whenever there is a need for determining an optimum order (sequence) of performing a number of jobs by a number of facilities according to some pre-assigned order so as to optimize the output in terms of time, cost or profit. Some practical situations in which knowledge of an optimum order of doing things is always desired are listed below: Patients waiting in a hospital for different types of tests; a number of items being processed through a finite number of machines, etc.

The *general sequencing problem* may be described as: Let there be n jobs to be performed one at a time on each ofw machines. The sequence (order) of the machines in which each job should be performed is given. The actual or expected time required by the jobs on each of the machines is also given. The general sequencing problem, therefore, is to" find the sequence out of $(n!)^m$ possible sequences which minimize the total elapsed time between the start of the job in the first machine and the completion of the last job on the last machine.

In particular, if n = 3 and m = 3, then total number of possible sequences will be $(3!)^3 = 216$. Theoretically, it may be possible to find optimum sequence but it will require a big computational time. Thus, one should adopt sequencing technique.

To find optimum sequence we first calculate the total elapsed time for each of the possible sequences. As stated earlier, even if values of m and n are very small, it is difficult to get the desired sequence with total minimum elapsed time. However, due to certain rules designed by Johnson, the task of determining an optimum sequence has become quite easy.

Sequencing problems are generally involved in production sections where the materials flow through a succession of processes to get converted into finished products. In general, the main objective of the production section is to make more efficient use of available facilities resulting into greater output. Work scheduling assigns work to each centre and determines the time schedule for each operation with efficient utilization of facilities and, thus, is an important function of the company.

In order to make efficient utilization of facilities, following should be considered:

(i) Keeping waiting time due to non-availability of work to a minimum.

(ii) Sufficient work should be provided from company's sales department.

(iii) Production control department of the company should ensure smooth flow of work through different work centres.

(iv) It must be ensured that work departments' facilities do not remain idle for need of work from the preceding work departments.

(v) Sequencing of the jobs to the facilities should be planned such that the facilities at work departments are efficiently utilized producing the jobs in least possible time.

Though the approach of "first come first scheduled" makes the scheduling process simpler, it does not provide the economy in production. Therefore, an approach is required to improve utilization of available facilities by improving decision rules relating to the sequencing of work over these facilities. Sequencing is concerned with a situation where measure of efficiency is a function of the order (sequence) in which a series of tasks are carried out. Sequencing models serve as an ideal tool for finding the optimal sequence of jobs and help the management in determining the optimal job sequence. The efficiency of the technique increases with backlog of orders. However, the limitation of the technique is that all jobs should have a definite operation sequence. The sequencing problems can be put into following two groups:

(i) Each of the tasks requiring processing on some or all of different machines, in which the effectiveness for any given order of the jobs (tasks) at each machine can be measured and be selected from the $(n!)^m$ theoretically possible sequences as following:

 (a) Technologically feasible sequences satisfying the restrictions (if any) on the sequence on which task has to be worked through m machines.

 (b) One or several of the technologically feasible sequences leading to optimum measure of efficiency.

 Though, it is possible to determine the best sequence by enumeration, in practice, the computation for the given sequence may be quite complex and time consuming. In a number of situations, this approach may be prohibitive even for moderate values of m and n. Even if the number of tasks to be worked are limited (say *n*) and even operation sequence is fixed, the number of alternative sequences will be large, i.e., *n*!. For example, consider the case in which 4 jobs are to be worked, each requiring time on each of 3 machines. There can be $(4!)^3$ or 14,824 different sequences. However, some of these sequences may not be feasible due to the requirement of certain operations to be carried out in a particular sequence.

(ii) In the second type of problem, there is a job shop with a number of machines and a list of jobs to be worked. In this case, when a machine completes the job, decision is to be made about the next job to be worked by this machine. The main characteristic of the situation is that the list of jobs will go on changing with fresh orders received.

NOTATIONS, TERMINOLOGY AND ASSUMPTIONS

Notations

t_{ij} = Processing time (time required) for job i on machine j.

T = Total elapsed time for processing all the jobs. This includes idle time, if any.

I_{ij} = Idle time on machine j from the end of job (i – 1) to the start of job i.

Terminology

- *Number of* Machines The number of machines refer to the number of service facilities through which a job must pass before it is assumed to be completed.
- *Processing Time* It is the time required by a job on each machine.
- *Processing Order* It refers to the order (sequence) in which machines are required for completing the job.
- *Idle Time on a Machine* It is the time for which a machine does not have a job to process, i.e. idle time from the end of job (i – 1) to the start of job i.
- *Total Elapsed Time* It is the time interval between starting the first job and completing the last job including the idle time (if any) in a particular order by the given set of machines.

No *Passing Rule* It refers to the rule of maintaining the order in which jobs are to be processed on given machines. For example, if n jobs are to be processed on two machines M_1 and M_2 in the order M_1,M_2, then each job should go to machine M_1, first and then to M_2.

Assumptions

1. The processing time on different machines are exactly known and are independent of the order of the jobs in which they are to be processed.
2. The time taken by the job in moving from one machine to another is negligible.
3. Once a job has begun on a machine, it must be completed before another job can begin on the same machine.
4. All jobs are known and are ready for processing before the period under consideration begins.
5. Only one job can be processed on a given machine at a time.
6. Machines to be used are of different types.
7. The order of completion of jobs are independent of the sequence of jobs.

GENERAL SEQUENCING PROBLEM

Let there are n jobs (1, 2, 3,....n), each of which has to be processed one at a time on each of m machines (A, B, C,......). Following are known/given:

(a) Orders of processing each job through machines, e.g., job-1 is to be processed through machines A, B, C.

(b) Actual or expected time required by the jobs on each machine

The problem is to determine a sequence out of $(n!)^m$ number of all possible sequences (alternatives) for processing the jobs, which minimizes the total elapsed time for all jobs.

If

A_i = Time required for job i on machine A

B_i = Time required for job i on machine B

and so on

i = 1, 2, 3,....n

T = The total elapsed time (including the idle time of machine, if any)

Then the problem is to find for each machine a sequence of jobs. i.e., permutations of the integers 1, 2, 3,.....n (i_1, i_2, i_3.....i_n), which minimizes the time T.

Few examples of sequencing problems are:

(1) Computer centers having number of problems waiting for solution

(2) Emergency ward of hospital having several emergency cases waiting for the treatment

(3) Production units processing different products over various combinations of machines

The important assumptions made in solving the sequencing problem are:

(1) A task (job) is an entity though the task may represent a lot of individual parts, no lot can be processed by more than one machine at one time.

(2) At a time, only one operation can be carried out on each machine.

(3) Each operation, once commenced, is performed till completion.

(4) Each operation is to be finished before any other operation to be carried out in succession.

(5) There is only one number of each type of machine.

(6) Time intervals for processing do not depend on the order of operations.

(7) A task is processed earliest possible subject to sequencing requirements.

(8) All the tasks are known and ready for processing before the period under consideration begins.

(9) The time taken in transfer of tasks from machine to other machine is negligible.

It is not possible to solve all types of sequencing problems due to complexity of the system. The satisfactory solutions are available only in few cases. In the present chapter, following cases will be discussed:

1. *n* jobs to be processed on two machines (A and B) in the order AB
2. n jobs to be processed on 3 machines (A, B and C) in the order ABC
3. Two jobs and m machines
4. *n* jobs and *m* machines.

PROCESSING n JOBS THROUGH TWO MACHINES

Let there be jobs, each of which is to be processed through two machines, M_1 and M_2 in the order M_1M_2, i.e. each job has to be pass through the same sequence of operations. In other words, a job is assigned on machine M_1 first and after it has been completely processed on machine M_1 it is assigned to machine M_2. If the machine M_2 is not free at the moment for processing the same job, then the job has to wait in a waiting line for its turn on machine M_2 i.e. passing is not allowed.

Since passing is not allowed, therefore, machine M_1 will remain busy in processing all the njobs one-by-one while machine M_2 may remain idle time of the second machine. This can be achieved only by determining sequence of n jobs which are to be processed on two machines M_1 and M_2. The procedure suggested by Johnson for determining the optimal sequence can be summarized as follows.

The Algorithm

Step 1 List the jobs along with their processing times in a table as shown bleow:

Processing Time on Machine	*Job Number* 1	2	3	...	*n*
M_1	t_{11}	t_{12}	t_{13}	...	t_{1n}
M_2	t_{21}	t_{22}	t_{23}	...	t_{2n}

Step 2 Examine the columns for processing times on machines M_1 and M_2 and find the smallest processing time in each column, *i.e.* find out, min. (t_{1j}, t_{2j}) for all j.

Step 3(a) If the smallest processing time is for the first machine M_1 then place the corresponding job in the first available position in the sequence. If it is for the second machine, then place the corresponding job in the last available position in the sequence.

(b) If there is a tie in selecting the minimum of all the processing times, then there may be three situations:

(i) Minimum among all processing times is same for the machine, *i.e.* min $(t_{1j}, t_{2j}) = t_{1k} = t_{2r}$, then process the kth job first and the rth job last.

(ii) If the tie for the minimum occurs among processing times t_{1j} on machine M_1 only, then select the job corresponding to the smallest job subscript first.

(iii) If the tie for the minimum occurs among processing rimes t_{2j} on machine M_2, then select the job corresponding to the largest job corresponding to the largest job subscript last.

Step 4 Remove the assigned jobs from the table. If the table is empty, stop and go to Step 5. Otherwise, go to Step 2.

Step 5 Calculate idle time for machines M_1 and M_2:

(a) Idle time for machine M_1 = (Total elapsed time) – (Time when the last job in a sequence finishes on machine M_1)

(b) Idle time for machine M_2 = Time at which the first job in a sequence finishes on machine M_2) + $\sum_{j=2}^{n}$ {(Time when the jth job in a sequence starts on machine M_2) – (Time when the (j – l)th job in a sequence finishes on machine M_2)}.

Step 6 The total elapsed time to process all jobs through two machines is given by

Total elapsed time = Time when the nth job in a sequence finishes on machine M_2.

$$= \sum_{j=1}^{n} M_{2j} + \sum_{j=1}^{n} I_{2j}$$

where M_{2j} = Time required for processing yth job on machine M_2.

I_{2i} = time for whichmachine M_2 remains idle after processing (j – l)th job and before starting work in jth job.

The problem can be considered as:

(i) *n* jobs to be processed on 2 machines A and B,

(ii) Each job is processed in the order AB and no passing is allowed, i.e., which-ever job is processed second on machine A, should also be processed second on machine B, and so on. This situation is quite common in chemical industries and refineries where the material flows from work station to work station through pipes.

(iii) The exact or expected processing times A_1, A_2, A_3 ....A_n on machine A and B_1, B_2, B_3,......B_n on machine B are known.

The problem is to sequence the jobs so as to minimize the total elapsed time T. The solution procedure is given by Johnson in 1954 represents the sequence of jobs (1, 2, 3,......n) on machines A and B. Job-1 occupies machine A for A_1 hours, while machine B is idle for time X_1. As soon as job-1 is completed on machine A, job-2 goes on it, and job-1 goes to machine B and so on. The total elapsed time T is obtained by the point of time at which job-1 enters on machine-A and the point of time at which job-n is completed on machine B. At any time, machine B is either idle or working. Therefore

$$T = \sum_{i=1}^{n} B_i + \sum_{i=1}^{n} X_i$$

Where some of the B_i's may be zero.

Since $\sum_{i=1}^{n} X_i$ is the total time for which machine B has to work and thus, being constant, does not form a part of minimizing T. Therefore, the problem is that of minimizing $\sum_{i=1}^{n} X_i$.

It can be observed that

$$X_1 = A_1$$

$$X_2 = \begin{cases} A_1 + A_2 - X_1 - B_1, & \text{if } A_1 + A_2 > X_1 + B_1 \\ 0 & \text{otherwise} \end{cases}$$

The above expression for X_2 can be given as

$$X_2 = \max\,[A_1 + A_2 - X_1 - B_1,\ 0]$$

Thus $X_1 + X_2 = \max\,[A_1 + A_2 - B_1,\ A_1]$ since $X_1 = A_1$

In the similar way

$$X_3 = \max\,[A_1 + A_2 + A_3 - B_1 - B_2 - X_1 - X_2,\ 0]$$

and $X_1 + X_2 + X_3$

$$= \text{Max}\left[\left(\sum_{i=1}^{3} A_i - \sum_{i=1}^{2} B_i\right), \sum_{i=1}^{2} X_i\right]$$

$$= \text{Max}\left[\left(\sum_{i=1}^{3} A_i - \sum_{i=1}^{2} B_i\right), \left(\sum_{i=1}^{2} A_i - B_i\right), A_1\right] \qquad (5.1)$$

In general, it can be given as,

$$\sum_{i=1}^{n} X_i = \text{Max}\left[\left(\sum_{i=1}^{n} A_i - \sum_{i=1}^{n-1} B_i\right), \left(\sum_{i=1}^{n-1} A_i - \sum_{i=1}^{n-2} B_i\right), A_1\right]$$

$$= \underset{1<r\leq n}{\text{Max}}\left[\left(\sum_{i=1}^{r} A_i - \sum_{i=1}^{r-1} B_i\right)\right] \qquad (5.2)$$

If $\sum_{i=1}^{n} X_i$ is denoted by $D_n(S)$, i.e., function of the sequence, S, the problem reduces to that of finding the sequence <S*> for processing the jobs 1, 2,.....*n* so as to have the inequality $D_n(S^*) \leq D_n(S_0)$ for any sequence <S_o> other than <S*>. Therefore, the expression (5.2) is evaluated for each positive value of *r* (1 through *n*) and maximum of all the values is taken as $D_n(S)$.

The solution procedure of sequencing *n* jobs on 2 machines can be summarized as below:

1. The least processing time occurring in the list A_i 's and B_i 's, i.e., A_1, A_2, A_3,........A_n and B_1, B_2, B_3,......B_n is chosen. If there is a tie, either of the smallest time is selected.
2. If the minimum processing time is A_k for some i = k, the kth job is processed first. If this minimum is B_r for some i = r, the rth job is done last. This decision applies to both machines A and B. If there is a tie for both minima $A_k = B_r$, kth job is processed first of all among the A_i 's, the job corresponding to the minimum of B_i 's is selected and it is processed first of all. If the tie for minimum occurs among the B_i 's, job corresponding to the minimum of A_i 's is selected and processed in the last.
3. The jobs already assigned are crossed out and steps 1 and 2 are repeated to sequence the jobs next to first or next to last.
4. Continue in the above manner until all jobs have been sequenced. The resulting sequence will give the minimum elapsed time, T.

PROCESSING n JOBS THROUGH THREE MACHINES A, B, & C

The problem can be considered as following:

(i) Three machines A, B, and C are involved in processing of n jobs.

(ii) Each job is performed on the three machines A, B and C in the order ABC.

(iii) Processing time of ith job on the three machines, i.e., A_i, B_i, C_i, respectively are known.

(iv) Transfer of jobs is not permitted, i.e., order over each machine is strictly followed.

So far, no general solution is available for such problems. However, by extending the earlier method adopted by Johnson (1954), special cases can be solved where either one or both of the following conditions hold.

(i) The least processing time on machine A ≥ maximum processing time on machine B, i.e.,

$$\text{Min } (A_i) \geq \text{Max } (B_j)$$

(ii) The least processing time on machine C ≥ maximum processing time on the machine B, i.e.,

$$\text{Min } C_i \geq \text{Max } (B_j)$$

The procedure stated here is without proof. This procedure replaces the problem with an equivalent problem involving n jobs and two fictitious machines represented by G and H. The corresponding processing times G_i and H_i are defined by

$$G_i = A_i + B_i \quad \text{and} \quad H_i = B_i + C_i$$

Now the sequence of jobs for these two machines G and H is determined in the usual manner, i.e., by applying the procedure for problems of n jobs and 2 machines as described above (Section-5.3). The resulting optimal solution will also be optimal to the original problem.

Johnson provides an extension of his procedure to the case in which there are three instead of two machines. Each job is to be processed through three machines M_1, M_2 and M_3 in the order M_1 M_2 M_3. The list of jobs with their processing times is given below. An optimal solution to this problem can be obtained if either or both of the following conditions hold good:

Processing Time on Machine	***Job Number***				
	1	***2***	***3***	...	***n***
M_1	t_{11}	t_{12}	t_{13}	...	t_{1n}
M_2	t_{21}	t_{22}	t_{23}	...	t_{2n}
M_3	t_{31}	t_{32}	t_{33}	...	t_{3n}

1. The minimum processing time on machine M_1 is at least as great as the maximum processing time on machine M_2, that is,

$$\min t_{ij} \geq \max t_{2j}, \qquad j = 1, 2, ..., n$$

2. The minimum processing time on machine M_3 is at least as great as the maximum processing time on machine M_2, that is,

$$\min t_{3j} \geq \max t_{2j}, \qquad j = 1, 2, ..., n$$

If either or both the above conditions hold good, then the steps of the algorithm can be summarized in the following steps.

The Algorithm

Step 1 Examine processing times of given jobs on all three machines and if either one or both the above conditions hold, then go to Step 2, otherwise the algorithm fails.

Step 2 Introduce two fictitious machines, say G and H with corresponding processing times given by

(i) $t_{Gj} = t_{1j} + t_{2j}, \quad j = 1, 2, ..., n$

that is, processing time on machine G is the sum of the processing times on machines M_1 and M_2 and

(ii) $t_{Hj} = t_{2j} + t_{3j}$,
$j = 1, 2, ..., n$

that is, processing time on machine H is the sum of the processing times on machines M_2 and M_3.

Step 3 Determine the optimal sequence of jobs for this n-job, two machine equivalent sequencing problem with the prescribed ordering GH in the same way as discussed earlier.

PROCESSING n JOBS THROUGH m MACHINES

Let there be n jobs, each of which is to be processed through m machines, say M_1, M_2, ..., M_m in the order M_1 M_2... M_3. The optimal solution to this problem can be obtained if either or both of the following conditions hold good.

(a) Min $\{t_{1j}\} \geq$ Max $\{t_{ij}\}$; $j = 2, 3, \ldots, m - 1$

and/or (b) Min $\{t_{mj}\} \geq$ Max $\{t_{ij}\}$; $j = 2, 3, ..., m - 1$

that is, the minimum processing time on machines M_1 and M_m is as great as the maximum processing time on any of the remaining $(m - 1)$ machines.

If either or both these conditions hold good, then the steps of the algorithm can be summarized in the following steps.

Step 1 Find, Min $\{t_{1j}\}$, Min $\{t_{mj}\}$ and Max $\{t_{ij}\}$ and verify above conditions. If either or both the conditions mentioned above hold, then go to Step 2. Otherwise the algorithm fails.

Step 2 Convert w-machine problem into 2-machine problem by introducing two fictitious machines, say G and H with corresponding processing times given by

(i) $t_{Gj} = t_{1j} + t_{2j} +... + t_{m-1j}$; $j = 1, 2, ..., n$

i.e. processing time of n-jobs on machine G is the sum of the processing times on machines M_1, M_2, ..., m_{m-1}, j

(ii) $t_{Hj} = t_{2j} + t_{3j} +... + t_{mj}$; $j = 1, 2, ..., n$

i.e. processing time ofn-jobs on machine H is the sum of the processing times on machines M_2, M_3,..., M_m.

Step 3 The new processing times so obtained can now be used for solving n-job, two-machine equivalent sequencing problem with the prescribed ordering HG in the same way as discussed earlier.

Remarks 1. In addition to the conditions given in Step 2, if

$$t_{2j} + t_{3j} + ... + t_{m-1j} = k \text{ (constant)}$$

for ally = 1, 2, ..., m – 1, then the optimal sequence can be obtained for n-jobs and two machines M_1 and M_m in the order M_1M_m as usual.

2. If t_{1j}, = t_{mj} and t_{Gj} = t_{Hj}, for all j = 1, 2, ..., n, then total number of optimal sequences will be n and total minimum elapsed time in these cases would also be the same.

3. The method described above for solving n-jobs and w-machines sequencing problem is not a general method. It is applicable only to certain problems where the minimum cost (or time) of processing the jobs through first and/or last machine is more than or equal to the cost (or time) of processing the jobs through remaining machines.

n JOBS THROUGH *m* MACHINES

The rule used in processing n jobs through 3 machines can be generalized for n jobs through m machines (M_1, M_2, M_3, ... M_n) in the order M_1, M_2, M_3,............M_n under some conditions given below:

if a_{ij} = processing time of *i*th job on *jth* machine

$i = 1, 2, \ldots\ldots\ldots\ldots n,$

and $j = 1, 2, \ldots\ldots\ldots\ldots m$

Minimum processing times for the first and the last machines and maximum processing times for all intermediate machines are calculated, i.e.,

$\min\limits_i a_{i1}$, $\min\limits_i a_{im}$ and $\max\limits_i a_{ij}$ are determined.

The problem can be solved only if either of the following two or both conditions are fulfilled.

(i) Minimum time or cost of processing the jobs through first machine is greater than or equal to the time or cost of processing the jobs through intermediate machines, i.e.

$$\text{Min } a_{i1} \geq \text{Max } a_{ij}, \text{ for all } j = 2, 3, \ldots m-1$$

(ii) Minimum time or cost of processing the jobs through last machine is greater than or equal to the time or cost of processing the jobs through intermediate machines, i.e.,

$$\text{Min } \alpha_{im} \geq \text{Max } \alpha_{ij}, \quad \text{for all } j = 2, 3, \ldots m-1$$

If atleast one of the above two conditions is fulfilled, the method can be adopted as following:

(i) m machine problem is converted into 2 machine problem considering two fictitious machines G and H with their processing times G_i and H_i, respectively. Then

$$G_i = \alpha_{i1} + \alpha_{i2} + \ldots\ldots\ldots\ldots + \alpha_{im-1}{}',\ i = 1, 2, \ldots\ldots n$$

$$H_i = \alpha_{i2} + \alpha_{i3} + \ldots\ldots\ldots\ldots + \alpha_{im}{}',\ i = 1, 2, \ldots\ldots n$$

(ii) The optimal sequence of *n* jobs through 2 machines is determined by using the optimal sequence algorithm as in case of *n* jobs through 3 machines. The sequence, which is optimal for *G* and *H* machines, will also be optimal for the original problem.

(iii) If $\alpha_{i2} + \alpha_{i3} + \ldots\ldots\ldots + \alpha_{im-1}$ = Constant (*C*) where *C* is a fixed positive number for all $i = 1, 2, \ldots\ldots n$, then the required optimal solution can be determined by solving the problem for *n* jobs and 2 machines M_1 and M_m in the order M_1 M_m.

PROCESSING TWO JOBS THROUGH *m* MACHINES

This deals with the problem involving following:

(i) Only two jobs, say 1 and 2 to be processed

(ii) *m* machines, say A, B, C,........... M.

(iii) Exact of expected processing times of jobs on the machines, A_1, A_2, B_1, B_2,............. M_1, M_2 are known.

(iv) The technological sequencing of machines for each job is known in advance.

The problem will be to determine the sequence of jobs for machines which minimizes the total elapsed time from the start of the first job to the completion of the last job. Though there are 2^m possible sequences, all will not be technologically feasible. Therefore, it will require for deleting the infeasible sequences. Furthermore, among the feasible sequences, many of them may not be optimum and, thus, it is essential to identify them. There are two techniques to solve such problems, i.e.,

1. Graphical Method
2. Non-numerical approach developed by Akers and Friedman

Processing Two Jobs Through m Machines

Let there be two jobs A and B each of which is to be processed on m machines say M_1, M_2,..., M_m in two different orders. The technological ordering of each of the two jobs through m machines is known in advance. Such ordering may not be same for both the jobs. The exact or expected processing times on the given machines are known. Each machine can perform only one job at a time. The objective is to determine an optimal sequence of processing the jobs so as to minimize total elapsed time.

The optimal sequence in this case can be obtained by using graph. The procedure can be illustrated by taking examples.

SOLVED EXAMPLES

Example 1:

Find the sequence that minimizes the total elapsed time required to complete the following tasks on two machines.

Task	:	*A*	*B*	*C*	*D*	*E*	*F*	*G*	*H*	*I*
Machine I	:	*2*	*5*	*4*	*9*	*6*	*8*	*7*	*5*	*4*
Machine II	:	*6*	*8*	*7*	*4*	*3*	*9*	*3*	*8*	*11*

Solution:

The smallest processing time between the (wo machines is 2 which corresponds to task A on machine I. Thus, task A will be processed first as shown below:

A							

After the task A has been set for processing first, we are left with 8 tasks and their processing times as given below:

Task	:	B	C	D	E	F	G	H	I
Machine I	:	5	4	9	6	8	7	5	4
Machine II	:	8	7	4	3	9	3	8	11

The minimum processing time in this reduced problem is 3 which corresponds to task E and G both on machine II. Since the corresponding processing time of task E on machine I is less than the corresponding processing time of task G on machine I, therefore, task E will be processed in the last and task G shall be processed before it. Tasks E and G will not be considered further. Thus, current partial assignment becomes:

A						G	E

The set of processing times now gets reduced to:

Task	:	B	C	D	F	H	I
Machine I	:	5	4	9	8	5	4
Machine II	:	8	7	4	9	8	11

The smallest processing time in this reduced problem is 4, which corresponds to task C and I on machine I and to task D on machine II. Thus task C will be placed in the second sequence cell and task I in the third sequence cell and task D in the sequence cell before task G. The entries of the partial sequence are now:

A	C	I				D	G	E

The set of processing times now gets reduced as follows:

Task	:	B	F	H
Machine I	:	5	8	5
Machine II	:	8	9	8

The smallest processing time in this reduced problem is 5, which corresponds to tasks B and H both on machine I. Since the corresponding processing times of B and H on machine II is same, therefore, either of these two tasks can be placed in fourth and fifth sequence cells. Thus, it indicates an alternative optimal sequence. The optimal sequences are, therefore, given below:

A	C	I	B	H	F	D	G	E

A	C	I	H	B	F	D	G	E

The minimum elapsed time for machines I and II is calculated as shown in Table 5.1.

Table 5.1 : Minimum Elapsed Time

Task Sequence	*Machine I*		*Machine II*	
	Time In	*Time Out*	*Time In*	*Time Out*
A	0	2	2	8
C	2	6	8	15
I	6	10	15	26
B	10	15	26	34
H	15	20	34	42
F	20	28	42	51
D	28	37	51	55
G	37	44	55	58
E	44	50	58	61

In Table 5.1 the minimum elapsed time, i.e. time from start of task A to completion of last task E is 61 hours. During this time the machine I remains idle for 61 – 50 = 11 hours. The idle time for machine II is equal to the time at which the first task A in the sequence finishes on machine I plus the last task E in the sequence starts on machine II minus the last but one task G finishes on machine II, i.e. 2 + 58 – 58 = 2 hours.

Example 2:

There are five jobs, each of which must go through the two machines A and B in the order AB. Processing times are given below:

Processing Time (hours)

Job	*1*	*2*	*3*	*4*	*5*
Machine A	*5*	*1*	*9*	*3*	*10*
Machine B	*2*	*6*	*7*	*8*	*4*

Determine a sequence for 5 jobs that will minimize the elapsed time T.

Solution:

Applying the solution procedure as given above, it is found that the least processing time is 1 hour for job-2 on the machine *A*. As a result, job-2 will be listed at first place as shown below:

2			

The reduced list of processing times then becomes

Processing Time (hours)

Job	***1***	***3***	***4***	***5***
Machine A	5	9	3	10
Machine B	(2)	7	8	4

Again the least processing time in the above reduced list is 2 hours for job-1 on the machine B. Thus, job-1 is processed last and can be listed as below:

2			1

Continuing in the identical manner, the next reduced list would be

Processing Time (hours)

Job	***3***	***4***	***5***
Machine A	9	(3)	10
Machine B	7	8	4

This will result in the following sequence and list

2	4			1

Sequence

	Processing Time (hours)	
Job	***3***	***5***
Machine A	9	10
Machine B	7	(4)

This will lead to the sequence

2	4			5	1

Thus finally the optimal sequence is obtained as

2	4	3	5	1

Now it is possible to compute the minimum elapsed time corresponding to the optimal sequencing using the individual processing time given in the statement of the problem. The details can be given either on a Gantt Chart tabular form (Table 5.2).

Table 5.2

Job Sequence	*Machine A*		*Machine B*	
	Time in	*Time out*	*Time in*	*Time out*
2	0	1	1	7
4	1	4	7	15
3	4	13	15	22
5	13	23	23	27
1	23	28	28	(30)

Table 5.2, it can be observed that the minimum elapsed time is 30 hours. During this time machine A remains idle for 2 hours (from 28 to 30 hours) and the machine B remains idle for 3 hours only (from 0-1, 22-23 and 27-28 hours).

In the present problem, it may be observed that job may be held in inventory before going to the machine, *e.g.,* 4th job will be free on machine-A after 7th hour. Thus, it will be kept in inventory for 3 hours. In this problem, it is assumed that storage space is available and cost of holding the inventory for each job is either same or negligible. The second general assumption is that the sequence of completion of jobs has no significance, i.e., no priority for jobs is claimed.

Example 3:

Solve the following sequencing problem giving an optimal solution when passing is not allowed.

Machine			*Job*		
	A	*B*	*O*	*D*	*E*
M_1	*11*	*13*	*9*	*16*	*17*
M_2	*4*	*3*	*5*	*2*	*6*
M_3	*6*	*7*	*5*	*8*	*4*
M_4	*15*	*8*	*13*	*9*	*11*

Solution:

From the data of the problem it is observed that

$$\text{Min}\left(t_{M_1,j}\right) = 9 = t_{M_1,C;} \quad \text{Min}\left(t_{M_4,j}\right) = 8 = t_{M_4,B}$$

and
$$\text{Min}\left\{t_{M_2,j}\right\} = 6 = t_{M_2,E;} \quad \text{Min}\left\{t_{M_3,j}\right\} = 8 = t_{M_3,D}$$

Since both the conditions

$$\text{Min}\left\{t_{M_1,j}\right\} \geq \text{Max}\left\{t_{M_2,j};\, t_{M_3,j}\right\}$$

$$\text{Min}\left\{t_{M_4,j}\right\} \geq \text{Max}\left\{t_{M_2,j};\, t_{M_3,j}\right\};\ j = 1, 2, ..., 5$$

are satisfied, therefore given problem can be converted into a 5-jobs and 2-machine problem as G and H.

Further, it may be noted that, Min $t_{M_2,j} + t_{M_3,j} = 10$ (a fixed constant) for all j (j = 1, 2,.... 5). Thus the given problem is reduced to a problem of solving 5-jobs through 2-machines M_1 and M_4 in the order M_1 M_4. This means machines M_2 and M_4 will have no effect on the optimality of the sequences.

The processing times of 5 jobs on machine M, and M, is as follows:

Job	:	A	B	C	D	E
Machine M_1	:	11	13	9	16	17
Machine M_2	:	15	8	13	9	11

Now using the algorithm described earlier, the optimal sequence so obtained is as follows:

C	A	E	D	B

The total elapsed time corresponding to the optimal sequence is 83 hours as shown in Table 5.3, using the individual processing times given in the original problem:

Table 5.3 : Minimum Total Elapsed Time

Job		*Machine*		
	M_1	M_2	M_3	M_4
C	0-9	9-14	14-19	19-32
A	9-20	20-24	24-30	32-45
E	29-36	36-42	42-46	46-57
D	36-52	52-54	54-62	62-71
B	52-65	65-68	68-75	75-83

Example 4:

Two jobs are processed on 5 machines as shown below:

Job-1 Sequence of machine times	*A*	*B*	*C*	*D*	*E*
	2	*3*	*4*	*6*	*2*
Job-2 Sequence of machine times	*C*	*A*	*D*	*E*	*B*
	4	*5*	*3*	*2*	*6*

Using graphical method, determine the minimum time needed to process the jobs. Find the job, which should be done first. Also compute the total time required to complete both the jobs.

***Solution*:**

Following steps are followed to solve the problem of 2 jobs through *m* machines by graphical method.

1. First a set of axes is drawn. The horizontal axis represents the processing time on job-1, whereas the vertical axis represents the processing time on job-2.
2. Machine times for two jobs are placed on the corresponding axes in the given technological order. Machine A takes 2 hours and 5 hours for job-1 and job-2, respectively. This can be shown by constructing a rectangle PQRS for machine A. In the same way, other rectangles for other machines are constructed.

3. Programme is made by beginning from the origin and moving through various states of completion points until the "finish point" is obtained. Physical interpretation of the path, thus selected, will involve a series of segments being horizontal, vertical or diagonal with an angle of 45°. These represent as following:
   (a) Moving to the right indicates the processing of job-1 while job-2 is idle.
   (b) Moving upward represents the processing of job-2 while job-1 is idle.
   (c) Moving diagonally represents the simultaneous processing of both the jobs.

Since both the jobs cannot be processed simultaneously on the same machine, diagonal movement through the shaded area is not possible.

4. Optimal path is determined as following:
   (a) Optimal path minimizing idle time for job-1 is determined by horizontal movement.
   (b) optimal path minimizing idle time for job-2 is determined by vertical movement.
   (c) Path is so chosen on which diagonal movement is as high as possible.
5. The elapsed time is determined by adding the idle time for either of the jobs to the processing time for the job.

In the present problem, the idle time for the selected path is observed to be 3 hours for the job-1, and zero for job-2. Therefore, the total time would be

$$2 + 3 + 4 + 6 + 2 + 3 = 20.$$

***Example 5*:**

There are two jobs to be processed through four machines A, B, C, and D in the following technological sequences:

Job-1	*A*	*B*	*C*	*D*
Job-2	*D*	*A*	*A*	*C*

Solution:

Processing times a_i, b_i, c_i, and d_i (i = 1, 2,.....) are as

Job/Machine	***A(a$_i$)***	***B(b$_i$)***	***C(c$_i$)***	***D(d$_i$)***
1	2	4	3	2
2	3	5	6	3

Obviously the choice is from two sequences of jobs for each machine, i.e., (2, 2) or (2, 1). Thus, one sequence is to be selected out of these two for each machine. Therefore, at time zero, either job-1 on machine A or job-2 on machine D is processed. It may be possible to gain something by beginning one job first while keeping the other machine idle for sometime. Similar situations may happen at other stages of the schedule, i.e., should the job be started through its next process soon after the machine becomes available, or should it wait? Therefore, the sequencing problem is to find for each machine whether the jobs should be processed 1, 2, or 2, 1?

In case of general problem, the set of *m* decisions would be called a program. In the present problem, there are only $2^4 = 16$ programs. Thus, the problem is to select the program out of 16, which gives the minimum elapsed time. The technique is ultimately based on enumeration of each program. However, large number of programs is a great hinderance in solving the problem and, thus, some programs have to be discarded. This can be carried out as following:

Firstly, some of these programs are deleted being infeasible on the basis of given technological ordering of the machines. Therefore, first of all, all the 16 programs are written using the following symbols:

a = Decision of job-1 before job-2 on machine A.

$\bar{a}$ = Decision of job-2 before job-1 on machine A

and similarly b, $\bar{b}$, c, $\bar{c}$ and d, $\bar{d}$ for other machines

In these notations, following table (Table 5.4) is obtained showing the 16 possible programs.

Tabel 5.4

Program No.															
1	2	3	4	5	6	7	8	9	10	11	12	13	14	15	16
$\bar{a}$	a	$\bar{a}$	a	$\bar{a}$	a	$\bar{a}$	a	$\bar{a}$	a	$\bar{a}$	a	$\bar{a}$	a	$\bar{a}$	a
$\bar{b}$	$\bar{b}$	b	b	$\bar{b}$	$\bar{b}$	b	b	$\bar{b}$	$\bar{b}$	b	b	$\bar{b}$	$\bar{b}$	b	b
$\bar{c}$	$\bar{c}$	$\bar{c}$	$\bar{c}$	c	c	c	c	$\bar{c}$	$\bar{c}$	$\bar{c}$	$\bar{c}$	c	c	c	c
$\bar{d}$	$\bar{d}$	$\bar{d}$	$\bar{d}$	$\bar{d}$	$\bar{d}$	$\bar{d}$	$\bar{d}$	d	d	d	d	d	d	d	d

It may be noted that the programmes give the ordering of the two jobs on a given machine but do not indicate any thing about the technological ordering of the machines for a given job. Therefore, $\bar{a}\ \bar{b}\ \bar{c}\ \bar{d}$ (Program-1) just indicates that "job-2 is to be done on each machine before job-1".

For two jobs to be processed on *m* machines with given technological orderings where X must precede machine Y for job–1, and X follows Y for job-2 (roughly X and Y happen in reverse orders in two routings), then the feasible program does not involve the term $\overline{X}\,Y$. This rule can be proved easily as below:

Technologically, following relationships should exist between the events:

(1 on X) preceds (1 on Y) (i)

(2 on Y) preceds (2 on X) (ii)

A program involving $\overline{X}\,Y$ represents

(2 on X) preceds (1 on X) (iii)

(1 on Y) preceds (2 on Y) (iv)

Now according to the statements (i), (iv), and (ii)

(1 on X) preceds (1 on Y) preceds (2 on Y) preceds (2 on X) or finally (1 on X) preceds (2 on X) which violates the statement (iii). Therefore, the program containing $\overline{X}\,Y$ is not technologically feasible under the above circumstances.

In order to apply this rule, firstly all machine pairs appearing in reverse order for the two jobs have to be identified. In the given example, these machine pairs are:

AB, AD, BD, and CD

Therefore, all the programs containing $\overline{a}b$, $\overline{a}d$, $\overline{b}d$, and $\overline{c}d$ are not feasible technologically. Now deleting the unfeasible programs, i.e., 3, 7, 9, 10, 11, 12, 13, 14, and 15 from the chart of 16 programs, following program is obtained.

Program No.						
1	2	4	5	6	8	16
$\overline{a}$	a	a	$\overline{a}$	a	a	a
$\overline{b}$	$\overline{b}$	b	$\overline{b}$	$\overline{b}$	b	b
$\overline{c}$	$\overline{c}$	$\overline{c}$	c	c	c	c
$\overline{d}$	$\overline{d}$	$\overline{d}$	$\overline{d}$	$\overline{d}$	$\overline{d}$	d

Akers and Friedman generalized this rule in the form of a theorem and have shown that the number of feasible programs, say *N* on *m* machines is given by

$$N = 1 + m \sum_{k=2}^{m} i_k$$

Where,

i_k = number of times that k specific machines (k = 2, 3,.......m) occur in the same order in both the given technological ordering (while intervening machines are ignored).

For example, in the above case m = 4, k = 2, 3, 4

Therefore, i_2 = number of combinations of 2 machines occurring in the same order in both routings.

= 2 (as the machine A, C in order AC and B, C in the order occur in both routings)

Identically, $i_3 = 0$ and $i_4 = 0 \rightarrow N = 1 + 4 + 2 = 7$

In the subsequent step, some more programs from the remaining 7 feasible programs can be discarded on the basis of optimality. Such programs can be identified on the basis of following result, developed by Akers and Friedman.

"A feasible program for 2 jobs and *m* machines situation may be optimal if it has no free machines". A machine is said to be free, if the two jobs are processed on it consecutively and while it is working, other machines remain idle.

For example, in program-16, job-1 is to be done on machine D before job-2 and job-1 is to be done on machine D in the last, and hence, job-2 will start after the completion of job-1 on all machines. As a result, on machine D, two jobs go consecutively and when D operates, other machines remain idle. Hence, machine D is free. Obviously, this program involves much idle time and, thus, does not appear to be a good program.

Now, the programs, which are not optimal, are deleted from the reduced programs. The rules for this operation are as follows given in Table 5.5.

The dots indicate a probability of other machines, if any. In the present example, where technological orderings are ABCD and DBAC, it can be seen that rule-1 (i.e., A.. D Vs. D....) deletes program containing ad, and thus, program-16 is deleted. Making use of rule-2, (i.e., A . . . C Vs. . . AC. .), program-5 can be deleted which contains $\overline{ac}$. By rule-4 (i.e., .. BC Vs. .. B..C..), program-4 can be deleted which contains $b\overline{c}$. Rest of the rules are not applicable to the given technological orderings.

Table 5.5

Rule No.	*Technological Orderings*		*Delete Programmes*
	Job-1	*Job-2*	*Containing*
1	X . . . . Y	Y	XY
2	X . . . Y . . .	X . . Y . . .	$\bar{X}Y$
3	. . X Y . . .	. . XY . . .	$\bar{X}Y$
4	. . X Y . . .	. . X . . . Y . .	$X\bar{Y}$
5	. . X Y . . . Z . . .	. . . X . . . YZ . . .	$X\bar{Y}Z$
6	. . X . . . YZ . . .	. . XY . . Z . . .	$\bar{X}Y\bar{Z}$

Thus, there are following four program, which satisfy necessary conditions for both technological feasibility and the optimality:

Program No.	1	2	6	8
	$\bar{a}$	a	a	a
	$\bar{b}$	$\bar{b}$	$\bar{b}$	b
	$\bar{c}$	$\bar{c}$	c	c
	$\bar{d}$	$\bar{d}$	$\bar{d}$	$\bar{d}$

The final step is to compute the elapsed times for each of the four programs, and to select that program for which the elapsed time is the least.

Example 6:

Find the sequence that minimizes the total time required in preforming the following jobs on three machines in the order ABC. Processing times (in hours) are given in the following table:

Job	:	*1*	*2*	*3*	*4*	*5*
Machine A	:	*8*	*10*	*6*	*7*	*11*
Machine B	:	*5*	*6*	*2*	*3*	*4*
Machine C	:	*4*	*9*	*8*	*6*	*5*

Solution:

Here, min $(t_{Aj}) = 6$; min $(t_{Cj}) = 4$; max $(t_{Bj}) = 6$. Since min $(t_{Aj}) \geq (t_{Bj})$ for all j is satisfied, the given problem can be converted into a problem of 5 jobs and two machines. The processing time on two fictitious machines G and H can be determined by the following relationships:

$$t_{Gj} = t_{Aj} + t_{Bj}; \quad j = 1, 2, ..., n$$

and $$t_{Hj} = t_{Bj} + t_{Cj}; \quad j = 1, 2, ..., n$$

The processing times for the new problem are given below:

Job	:	1	2	3	4	5
Machine G	:	8+5 = 13	10+6 = 16	6+2 = 8	7+3 = 10	11+4 = 15
Machine H	:	5+4 = 9	6+9 = 15	2+8 = 10	3+6 = 9	4+5 = 9

When the algorithm described for n jobs on two machines is applied to this problem, the optimal sequence so obtained is given by:

3 2 5 1 4

The total minimum elapsed time is given in Table 5.5.

Table 5.5 : Minimum Elapsed Time

Job	*Machine A*		*Machine B*		*Machine C*	
Sequence	*Time In*	*Time Out*	*Time In*	*Time Out*	*Time In*	*Time Out*
3	0	6	6	8	8	16
2	6	16	16	22	22	31
5	16	27	27	31	31	36
1	27	35	35	40	40	44
4	35	42	42	45	45	51

Table 5.5 indicates that the minimum total elapsed time is 51 hours. The idle time for machines A, B and C is 9 (= 51 –42) hours, 6 (= 51 –45) hours and 9 {= (8 – 0) + (45 – 44)} hours, respectively.

Example 7:

Find the optimal sequence for processing 4 jobs A, B, C, D on four machines M_1, M_2, M_3 and M_4. Processing times are given below in Table 5.6.

Table 5.6

Job/Machine	*Processing Times α_{ij} in hours*			
	M (α_{i1})	*M (α_{i2})*	*M(α_{i3})*	*M(α_{i4})*
A	*12*	*2*	*10*	*12*
B	*15*	*5*	*4*	*14*
C	*16*	*0*	*3*	*19*
D	*13*	*3*	*6*	*15*

Solution:

From the above table (Table 5.6)

Minimum processing time on M_1, α_{i1} = 12

Minimum processing time on M_4, α_{i4} = 12

and for intermediate machines M_2 and M_3

Maximum processing time on M_2, (α_{i2}) = 5

Maximum processing time on M_3, (α_{i3}) = 10

Since min. α_{ij} > max, α_{i1} and max, α_{i3}, and both the conditions are fulfilled, the problem can be solved by converting the problem into a problem involving only two machines G and H with following processing times:

Jobs	***Processing Times*** G_i	H_i
A	12 + 2 + 10 = 24	2 + 10 + 12 = 24
B	15 + 5 + 4 = 24	5 + 4 + 12 = 21
C	16 + 0 + 3 = 19	0 + 3 + 19 = 22
D	13 + 3 + 6 = 22	3 + 6 + 15 = 24

Using the algorithm for solving the problem of *n* jobs and 2 machines, the optimal sequence obtained will be

	M_1	M_2	M_3	M_4
C	0-16	16-16	16-19	19-38
D	16 -29	29-32	32-38	38-53
A	29-41	41-43	43-53	53-65
B	41-56	56-61	61-65	65-79

Therefore, total elapsed time = 79 hours

Machine M_1 will be idle for 23 hours (79-56), M_2 will be idle for 69 hours (0-16, 16-29, 32-41, 43-56, 61-79), M_3 will remain idle for 56 hours (0-16, 19-32, 38-43, 53-61, 65-79) and M_4 for 19 hours (0-19).

Example 8:

Find an optimal sequence for the following sequencing problems of four jobs and five machines when passing is not allowed of which processing time (in hours) is given below:

Job	*Machine*				
	M_1	M_2	M_3	M_4	M_5
A	7	5	2	3	9
B	6	6	4	5	10
C	5	4	5	6	8
D	8	3	3	2	6

Also find the total elapsed time.

Solution:

Here, $\text{Min}\left(t_{M_1,j}\right) = 5 = t_{M_1,C}$

$$\text{Min}\left(t_{M_5,j}\right) = 6 = t_{M_5,D}$$

and $\text{Min}\left\{t_{M_2,j}, t_{M_3,j}, t_{M_4,j}\right. = \{6, 5, 6\}$ respectively.

Since the condition of $\text{Min}\left(t_{M_5,j}\right) \geq \text{Max}\left\{t_{M_2,j}, t_{M_3,j}, t_{M_4,j}\right\}$ is satisfied, therefore the given problem can be converted into a four jobs and two machines problem as G and H. The processing times of four jobs denoted by t_{Gj} and t_{Hj} on G and H, respectively are as follows:

Job	:	A	B	C	D
Machine G	:	17	21	20	16
Machine H	:	19	25	23	14

where $t_{Gj} = \sum_{i=1}^{m-1} t_{ij}$ and $t_{Hj} = \sum_{i=2}^{m} t_{ij}$

Now using the optimal sequence algorithm, the following optimal sequence can be obtained

A	C	B	D

The total elapsed time corresponding to the optimal sequence can be calculated as shown in Table 5.7, using the individual processing times given in the original problem:

Table 5.7 shows that the minimum total elapsed time is 51 hours. The idle time for machines M_1, M_2, M_3, M_4 and M_5 is 25, 33, 37 and 18 hrs, respectively.

Table 5.7 : Minimum Elapsed Time

Job Sequence	Machine				
	M_1	M_2	M_3	M_4	M_5
A	0-7	7-12	12-14	14-17	17-26
C	7-12	12-16	16-21	21-27	27-35
B	12-18	18-24	24-28	28-33	35-45
D	18-26	26-29	29-32	33-35	45-51

Example 9:

Determine the optimal sequence for 5 jobs each of which must go through 3 machines A, B, and C in the order ABC, that will minimize the elapsed time. The processing times are given below:

Job	*Processing Times*		
i	A_i	B_i	C_i
1	*8*	*5*	*4*
2	*10*	*6*	*9*
3	*6*	*2*	*8*
4	*7*	*3*	*6*
5	*11*	*4*	*5*

Solution:

In the problem

Min $A_i = 6$,

Max $B_i = 6$ and Min $C_i = 4$

Since one of the two conditions stated above is satisfied by Min A_i = Max B_i, the procedure adopted for *n* jobs and 2 machines can be adopted.

The equivalent problem becomes

Job	$G_i = (A_i + B_i)$	$H_i = (B_i + C_i)$
1	13	9
2	16	15
3	8	10
4	10	9
5	15	9

Because of ties, the possible optimal sequences are follows:

(i)	3	2	4	5	1
(ii)	3	2	1	5	4
(iii)	3	2	1	4	5
(iv)	3	2	5	4	1
(v)	3	2	5	1	4
(vi)	3	2	4	1	5

The minimum elapsed time for first sequence is shown in Table 5.8.

Table 5.8

Job	Machine A		Machine B		Machine C	
	Time in	Time out	Time in	Time out	Time in	Time out
3	0	6	6	8	8	16
2	6	16	16	22	22	31
4	16	23	23	26	26	32
5	23	34	34	38	38	43
1	34	42	42	47	47	51

Thus any of these ordering may be adopted to sequence the jobs through machines A, B, and C. They all will yield a minimum elapsed time of 51 hours. Idle time for machine A = 9 hours, for B = 31 hours, and for C = 19 hours.

Similarly, minimum elapsed time can be calculated for other sequences and found to be 51 hours in all the cases.

Example 10:

Use the graphical method to minimize the time needed to process the following jobs on the machines shown, i.e. each machine find the job which should be done first. Also calculate the total elapsed time to complete both jobs.

			<u>*Machine*</u>			
Job 1	*Sequence* :	*A*	*B*	*C*	*D*	*E*
	1 Time (hrs)	3	4	2	6	2

			<u>*Machine*</u>			
Job 2	*Sequence* :	*B*	*C*	*A*	*D*	*E*
	Time (hrs)	5	4	3	2	6

Solution:

The solution procedure for solving the above problem can be summarized in the following steps:

Step 1 : Draw the set of axes at right angle to each other where x-axis represents the processing time of job 1 on different machines while job 2 remains idle and y-axis represents processing time of job 2 while job 1 remains idle.

Step 2 : Mark the processing times for jobs 1 and 2 on x-axis and y-axis, respectively according to the given order of machines as shown in Fig. 5.1.

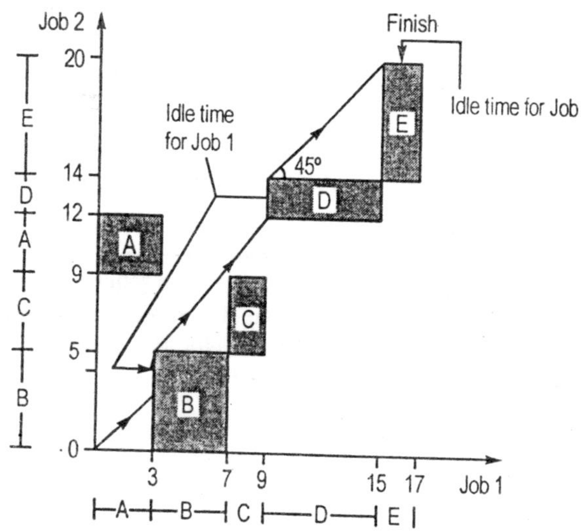

Fig. 5.1 : Graphical Solution of 2-Jobs and 2-Machines Sequencing Problem.

For example, machine A takes 3 hours for job 1 and 3 hours for job 2. Construct the rectangle for machine A as shown in Fig. 20.1. Similarly, construct other rectangles for machines B, C, D and E.

Step 3 Construct various blocks starting from the origin by pairing the same machines until a point marked 'finished' is obtained.

Step 4 Draw a line starting from origin to the point marked 'finish' by moving horizontally, vertically and diagonally along a line which makes an angle of 45° with the horizontal axis. Moving horizontally along this line indicates that first job is under process while second job is idle. Similarly, moving vertically along this line indicates that the second job is under process while first job is idle. The diagonal movement along this line shows that both the jobs are under process simultaneously.

Since simultaneous processing of both jobs on a machine is not possible, therefore, diagonal movement is not allowed. In other words, diagonal movement through rectangle areas is not allowed.

Step 5 An *optimal path* is one that minimizes the idle time for both the jobs. Thus, we must choose the path on which diagonal movement is maximum as shown in Fig. 20.1.

Step 6 Total elapsed time is obtained by adding the idle time for either job to the processing time for that job. In this example, the idle time for the chosen path is found to be 5 hrs and 2 hrs for jobs 1 and 2, respectively. The total elapsed time is calculated as follows:

Elapsed time. Job 1 = Processing time of job 1 + Idle time for job 1

= 17 + (2 + 3) = 22 hours

Elapsed time, Job 2 = Processing time of job 2 + Idle time for job 2

= 20 + (17 – 15) = 22 hours.

Example 11:

Using graphical method, calculate the minimum time needed to process jobs 1 and 2 on five machines A, B, C, D and E, i.e. for each machine find the job which should be done first. Also calculate the total needed to complete both jobs.

		<u>*Machine*</u>				
	Sequence :	A	B	C	D	E
Job	1 Time (hrs)	6	8	4	12	4
	Sequence :	B	C	A	D	E
Job 2	Time (hrs)	10	8	6	4	12

Solution:

Draw two axes at right angle to each other where x-axis represents the processing time of job 1 on different machines while job 2 remains idle and y-axis represents the processing time of job 2 on different machines while job 1 remains idle.

Mark the processing times for both the jobs 1 and 2 on x-axis and y-axis, respectively according to the given order of machines as shown in Fig. 5.2. For example, machine A takes 6 hours for job 1 as well as for job 2. Construct the rectangle for machine A as shown in Fig. 5.2. Similarly, construct other rectangles for machines B, C, D and E.

Draw a line starting from origin to the point marked 'finished' by moving horizontally, vertically and diagonally along a line which makes an angle of 45° with the horizontal axis. A horizontal move represents processing of job 1 while job 2 remains idle; a vertical move represents processing of job 2 while job 1 remains idle. The diagonal movement (i.e. a 45° line) shows that both the jobs are under process simultaneously. Since simultaneous orocessing of both the jobs on a machine is not possible, therefore, diagonal movement is not allowed.

An optimal path is one that minimizes idle time for both jobs 1 and 2. This means, this path should coincide with 45° line to the maximum extent. Thus, we must choose the path on which diagonal movement is maximum, as shown in Fig. 5.2. The total elapsed time is obtained by adding idle time for either job to the processing time for that job.

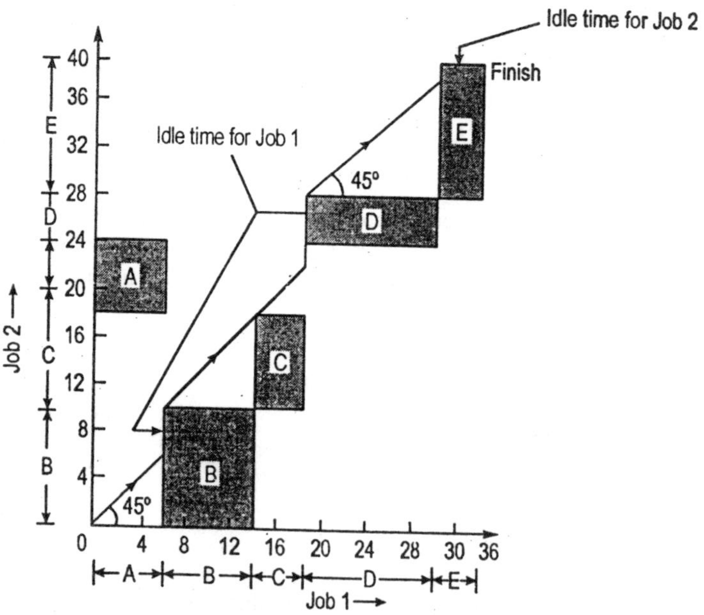

Fig. 5.2 : Graphical Solution of 2 Jobs and m Machines Sequencing Problem

The idle time for the chosen path is found to be 10 (= 4 + 6) hours for job 1 and 4 hours for job 2. The total elapsed time is calculated as follows:

Elapsed time. Job 1 = Processing time + Idle time = 34+(4+6) = 44 hours

Elapsed time. Job 2 = Processing time + Idle time = 40+(32–28)=44 hours.

EXERCISES

1. Explain briefly the solution procedure of processing of two jobs through the machines when the technological ordering of each of the jobs through the machines is prescribed in advance. Establish the following rule: If machine A precedes machine B for job I and machine B precedes machine A for job 2, then no programme that contains both the decisions: (a) job 2 before job 1, and (b)job 1 before job 2 on machine B, is technologically feasible.
2. What do you understand by the problem of sequencing? Discuss the various aspects of data required to formulate the problem of sequencing two jobs on m machines.
3. A job consists of N steps. Step i takes time t_i. If these jobs are grouped somehow into station systems, then twice as many units can be produced each day. Also two set-ups in parallel can also double the production rate. Critically examine the advantages of these two approaches.
4. What do you understand by the following terms in the context of sequence of jobs:
   (i) Job arrival pattern
   (ii) Number of machines
   (iii) The flow pattern in the shop
   (iv) The criteria of evaluating the performance of a schedule
5. Find an optimal sequence for the following sequencing problem of four jobs and five machines (when passing is not allowed) of which processing time (in hrs) is as follows;

Job	:	1	2	3	4
Machine M_1	:	6	5	4	1
Machine M_2	:	4	5	3	2
Machine M_3	:	1	3	4	2
Machine M_4	:	2	4	5	1
Machine M_5	:	8	9	7	5

   Also find the total elapsed time.
6. By using appropriate notion, obtain formulae for the following:
   (i) Waiting time of a job
   (ii) Completion time of a job

(iii) Mean flow time

(iv) Mean lateness

7. Five jobs have to be processed on the same machine. The set-up time for each job depends on the job processed earlier. A table of the set-up time is shown below. Find a sequence for processing all jobs that minimizes the total set-up costs.

Predecessor			***Follower Job***		
Job	***A***	***B***	***C***	***D***	***E***
A	0	29	20	18	24
B	0	0	14	19	16
C	0	35	0	37	26
D	0	15	10	0	10
E	0	18	16	40	0

8. Use the graphical method to find the minimum elapsed total time sequence of 2 jobs and 5 machines, when we are given the following information

				Machine		
	Sequence	:	A	B	C	D	E
Job	1 Time (hrs)	:	2	3	4	6	2
	Sequence	:	C	A	D	E	B
Job 2	Time (hrs)		4	5	3	2	6

9. We have five jobs, each of which must be processed on the two machines A and B in the order AB. Processing times in hours are given in the table below:

Job	:	1	2	3	4	5
Machined A	:	5	1	9	3	10
Machine B	:	2	6	7	8	4

Determine a sequence for the five jobs that will minimize the elapsed time T.

10. A book binder has one printing press, one binding machine, and manuscripts of a number of books. The time required to perform the printing and binding operations on each book are shown below. The binder wishes to determine the order in which the books should be processed, so that the total time required to process all books is minimized.

Book	:	1	2	3	4	5	6
Printing time (Hours)	:	30	120	50	20	90	110
Binding time (Hours)	:	80	100	90	60	30	10

11. Five jobs are performed, first on machine X and then on machine Y. The time taken, in hours by each job on each machine is given below:

Job	:	A	B	C	D	E
Time on machine A	:	12	4	20	14	22
Time on machine Y	:	6	14	16	18	10

Determine the optimum sequence of jobs that minimizes the total elapsed time to complete the jobs. Also compute the minimum time.

12. Following table shows the machine time (in hours) for 5 jobs to be processed on two different machines:

Job	:	1	2	3	4	5
Machined	:	3	7	4	5	7
Machined	:	6	2	7	3	4

Passing is not allowed. Find the optimal sequence in which jobs should be processed.

13. Find the sequence that minimizes the total elapsed time and processing time in hours required to complete the follovvingjobs:

Job	:	1	2	3	4	5	6
Machined A	:	4	8	3	6	7	5
Machined B	:	6	33	7	2	8	4

14. We have six jobs, each of which must go through machines A, B and C in the order ABC. Processing time (in hours) are given in the following table:

Job	:	1	2	3	4	5	6
Machined A	:	8	3	7	2	5	1
Machine B	:	3	4	5	2	1	6
Machine C	:	8	7	6	9	10	9

Determine a sequence for the five jobs that will minimize the elapsed time t.

15. Determine the optimal sequence of jobs that minimize the total elapsed time based on the following information. Processing time on machines is given in hours and passing is not allowed.

Job	:	A	B	C	D	E	F	G
Machine (M_1)	:	3	8	7	4	9	8	7
Machine (M_2)	:	4	3	2	5	1	4	3
Machine (M_3)	:	6	7	5	11	5	6	12

16. We have five jobs each of which must go through the machines A, B and C in the order ABC. Processing times (in hours) is as follows:

Job	:	1	2	3	4	5
Machine A	:	5	7	6	9	5
Machine B	:	2	1	4	5	3
Machine C	:	3	7	5	6	7

Determine the sequence for the jobs that will minimize the total elapsed time.

17. Find the sequence that minimizes the total elapsed time required to complete the following tasks. Each job is processed in the order ABC.

Job	:	1	2	3	4	5	6	7
Machined A	:	12	6	5	11	5	7	6
Machined B	:	7	8	9	4	7	8	3
Machine C	:	3	4	1	5	2	3	4

18. Explain the four elements that characterize a sequencing problem.
19. Explain the principal assumptions made while dealing with sequencing problems.
20. Give Johnson's procedure for determining an optimal sequence for processing n items on two machines. Give justification of the rule used in the procedure.
21. What is no passing rule in a sequencing algorithm? Explain the principal assumptions made while dealing with sequencing problems.
22. Give three different examples of sequencing problems from your daily life.
23. Write a short note on the 'sequencing decision problem for n jobs on two machines'.
24. Give a mathematical formulation of the optimal assignment and travelling salesman problems. In what way do the feasible solution matrices in the two cases differ?
25. A machine operator has to perform three operations, turning, threading and knurling on a number of different jobs. The time required to perform these operations (in minutes) for each job is known.

Job	*Time for Turning (minutes)*	*Time for Threading (minutes)*	*Time for Knurling (minutes)*
1.	3	8	13
2.	12	6	14
3.	5	4	9
4.	2	6	12
5.	9	3	8
6.	II	1	13

Determine the order in which the jobs should be processed in order to minimize the total time required to turn out all the jobs.

26. A ready-made garment manufacturer has to process 7 items through two stages of production, viz., cutting and sewing. The time taken for each of these items at the different stages are given below in appropriate units.

Item	:	1	2	3	4	5	6	7
Cutting time	:	5	1	3	4	6	7	12
Sewing time	:	2	6	7	5	9	5	8

(a) Find an order in which these items are to be processed through these stages so as to minimize the total processing time.

(b) Suppose a third stage of production is added, viz., pressing and packing, with processing time for these times as follows:

Item	:	1	2	3	4	5	6	7
Pressing and packing time	:	10	12	11	13	12	10	11

Find an order in which these 7 items are to be processed so as to minimize the time taken to process all the items through all the three stages.

27. Six jobs go first over machine I and then over II. The order of the completion of jobs has no significance. The following table gives the machine times in hours for six jobs and the two machines.

Job	:	1	2	3	4	5	6
Machine I	:	5	9	4	7	8	6
Machine II	:	7	4	8	3	9	5

Find the sequence of the jobs that minimizes the total elapsed time to complete the jobs. Find the minimum time by using Gantt Chart or by any other method.

28. We have five jobs each of which must go through two machines in the order AB. Processing times are given below:

Job	:	1	2	3	4	5
Machined A	:	10	2	18	6	20
Machine B	:	4	12	14	16	18

29. Two jobs are processed on 5 machines as shown below:

Job-1 Sequence of machine times	A	B	C	D	E
	2	3	4	6	2
Job-2 Sequence of machine times	C	A	D	E	B
	4	5	3	2	6

Using graphical method, determine the minimum time needed to process the jobs. Find the job, which should be done first. Also compute the total time required to complete both the jobs.

30. A manufacturing company processes 6 different jobs on two machines A and B. Number of units of each job and its processing times on A and B are given below:

Job	*Units of each Job*	*Processing*	*Time*
		Machine A (minutes)	*Machine B (minutes)*
1	3	5	8
2	4	16	7
3	2	6	11
4	5	3	5
5	2	9	7.5
6	3	6	14

Find the optimal sequence, the total minimum elapsed time and idle time for each machine.

31. Find the sequence that minimizes the total time required in preforming the following jobs on three machines in the order ABC. Processing times (in hours) are given in the following table:

Job	:	1	2	3	4	5
Machine A	:	8	10	6	7	11
Machine B	:	5	6	2	3	4
Machine C	:	4	9	8	6	5

32. We have seven jobs each of which has to go through the machine M_1 and M_2 in Ihe order M_1, M_2. Processing times (in hours) are given as:

Job	:	1	2	3	4	5	6	7
Machine A	:	3	12	15	6	10	11	9
Machine M B	:	8	10	10	6	12	1	3

Determine a sequence of these jobs that will minimize the total elapsed time T.

29. A machine operator has to perform two operations, turning and threading, on a number of different jobs. The time required to perform these operations (in minutes) for each job is known.

Job	*Time of Turning (minutes)*	*Time of Threading (minutes)*
1	3	8
2	12	10
3	5	9
4	2	6
5	9	3
6	11	1

Determine the order in which the jobs should be processed in order to minimize the total time required to turn out all the jobs.